Studies in Computational Intelligence

Volume 558

Series editor

Janusz Kacprzyk, Polish Academy of Sciences, Warsaw, Poland
e-mail: kacprzyk@ibspan.waw.pl

For further volumes:
http://www.springer.com/series/7092

About this Series

The series "Studies in Computational Intelligence" (SCI) publishes new developments and advances in the various areas of computational intelligence—quickly and with a high quality. The intent is to cover the theory, applications, and design methods of computational intelligence, as embedded in the fields of engineering, computer science, physics and life sciences, as well as the methodologies behind them. The series contains monographs, lecture notes and edited volumes in computational intelligence spanning the areas of neural networks, connectionist systems, genetic algorithms, evolutionary computation, artificial intelligence, cellular automata, self-organizing systems, soft computing, fuzzy systems, and hybrid intelligent systems. Of particular value to both the contributors and the readership are the short publication timeframe and the world-wide distribution, which enable both wide and rapid dissemination of research output.

Dmitry Gubanov · Nikolai Korgin
Dmitry Novikov · Alexander Raikov

E-Expertise: Modern Collective Intelligence

Springer

Dmitry Gubanov
Nikolai Korgin
Dmitry Novikov
Trapeznikov Institute of Control Science
Russian Academy of Sciences
Moscow
Russia

Alexander Raikov
Trapeznikov Institute of Control Science
Russian Academy of Sciences
Moscow
Russia

Russian Presidential Academy of National
   Economy and Public Administration
Moscow
Russia

and

Analytical Agency "New Strategy"
Moscow
Russia

ISSN 1860-949X   ISSN 1860-9503 (electronic)
ISBN 978-3-319-35715-7   ISBN 978-3-319-06770-4 (eBook)
DOI 10.1007/978-3-319-06770-4
Springer Cham Heidelberg New York Dordrecht London

Printed on acid-free paper

Springer is part of Springer Science+Business Media (www.springer.com)

Contents

Introduction

Today, the heads of federal and municipal authorities, the managers of corporations, industrial enterprises, and organizations (*Principals*) face a wide variety of aims and tasks. Furthermore, they operate under the complexity of existing problems, the diversity of subordinate departments and employees, dynamic or even uncertain requirements and conditions applied by an external environment. All these factors dictate that a Principal should have a subtle intuition, possess adequate information, and original techniques for efficient decision-making.

Any Principal disposes of limited time and resources including financial, cognitive, and intellectual resources. It seems often impossible for a Principal to comprehend a current situation in detail, to acquire and process all incoming information. Sometimes, a Principal finds difficulty in explaining his informational needs to subordinates. In other situations, full revelation of the Principal's aims may appear undesired, but the sense of discomfort causes inconvenience, and he needs competent advice. For hundreds or even thousands of years, a crucial role in management and decision-making has been performed by *expert* procedures. They proceed from acquiring and processing of the opinions of some *experts*, i.e., connoisseurs and specialists in corresponding subject domains. Historically, the term "expert" is associated with many other notions (an adviser, an assistant, a consultant, a foreteller, a prophet, a sage, a vizier, a master, an authority, a wizard, a pro, a dodger, a sorcerer, a magician, an extrasensory individual, a mentalist, to name a few).

Expertise represents both a field of scientific research [9, 24, 58, 80, 81, 91] and a field of practical activity. In the recent 50 years, thousands of expert organizations have been established worldwide. Since the early 2000s, this field demonstrates formation of a new phenomenon, *e-expertise*, which employs modern decision-making and information analysis technologies, data communication, and transmission networks. Due to the high complexity and responsibility of managerial decisions, a Principal frequently addresses the opinion of experts, professional communities, elite clubs, and influential public organizations. However, this leads to several effects, the so-called *expertocracy*. Following the development of informational community, we observe certain transformations in the existing management system; state authorities gradually become *transparent* (*responsibility management*), and management processes involve more and more subjects of civil society (*crowdsourcing*). In such conditions, decisions of state authorities are strongly dependent on expert assessments and opinions of civil society representatives. A Principal should consider the opinion of civil society (expressed by experts through dedicated information analysis technologies and network technologies). Experts legitimize lobby for the interests pursued by subjects of civil society including science intensive small- and medium-sized business companies.

The phenomenon of e-expertise gains particular importance in the context of distributed *situation centers*. They render technological, informational-analytical, and expert-analytical support of a Principal and his team in decision-making in various situations including unforeseen ones. Situation centers "compress" the period of decision-making by special methods of conducting meetings and organizing collective expert procedures with networked experts.

E-expertise bases on the following conventional definitions.

Expert activity—*Expertise* is a study of a certain object or subject, a situation, issue or topic, which requires special knowledge and results in a motivated report. Alternatively, expertise can be viewed as a method and process of assessing or identifying some properties, factors, obstacles, and tendencies in the development of a problem situation based on experts' involvement.

Expert appraisals (*assessments*) are judgments of high-level specialists and professionals in the form of content, qualitative, or quantitative estimation of an object or subject to-be-used in decision-making.

There exist *individual* and *collective* expert appraisals. Individual expert appraisals are generated by one highly skilled professional. For instance, a lecturer independently estimates the progress of a student, a physician diagnoses a patient. Nevertheless, in complicated cases (a nontrivial diagnosis or student expulsion), a collective opinion becomes vital—a council of physicians or a board of examiners solve the problem.

An expert is a specialist in a subject domain, a management connoisseur, a person informed of some event,

- possessing necessary flair, knowledge, and experience;
- being able to analyze and comprehend incoming information;
- being able to penetrate deeply into a problem situation and to assess a corresponding object or subject of expertise within his competence and awareness;
- handling necessary technologies, being able to assess their applicability in a specific situation of decision-making and provide appropriate recommendations and opinions;
- having definite rights and duties to a Principal;
- bearing personal responsibility for his opinions and recommendations.

An expert can be also defined as

- a subject (individual or collective) with knowledge, a personal opinion, and experience regarding a specific activity;

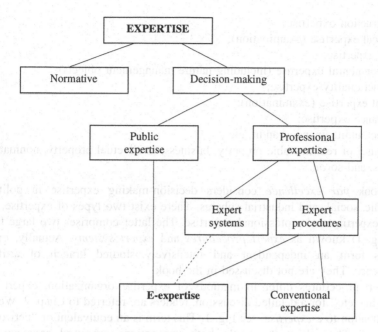

Fig. 1 E-expertise as a type of expertise

- a natural person having considerable intelligence in some field (politics, economics, social sphere, an industry or region, a science, a technology, public life);
- a legal entity performing a deep expert study of a given subject according to an established procedure;
- a representative of scientific and educational institutions, municipal authorities, public associations, and other organizations invited by a Principal as experts.

Subjects (individual or collective) of expert activity include a Principal, an analyst, a specialist in expert technologies, a coordinator of expertise, a moderator, a subject supervisor, and experts.

Similar to any expertise, e-expertise has several primary *aims*, namely

- increasing the sustainability and capitalization of a company, improving the quality of products and services, ensuring competitive ability;
- enhancing the reasonability level of decisions owing to the opinions of experts;
- controlling and/or establishing the correspondence between the characteristics of an expertise object or subject and requirements (conditions, restrictions) applied by normative legal documents of different levels.

According to the two primary aims of expertise, it seems possible to discriminate between *decision-making expertise* and *normative expertise* (see Fig. 1).

The following types of *normative expertise* are separated out depending on the fields of a specific professional activity:

- construction expertise;
- medical expertise (examination);
- legal expertise;
- environmental expertise (including nature management objects);
- product quality expertise;
- patent expertise (examination);
- insurance expertise;
- project audit, financial audit, etc.;
- appraisal of real/movable property, business, intellectual property, nonmaterial assets, and so on.

This book *par excellence* considers decision-making expertise in political, economic, social, and industrial spheres. There exist two types of expertise, *viz.*, public expertise and professional expertise. The latter comprises two large fields (see Fig. 1) known as *expert procedures* and *expert systems*. Actually, expert systems form an independent and intensively studied branch of artificial intelligence. They are not discussed in the book.

Expert procedures differ in methods of expertise organization, expert data processing, etc.; for a detailed discussion, readers are referred to Chap. 2. We pay most attention to *e-expertise*, see Fig. 1. This term is an equivalent of "networked expertise." Throughout the book, both conceptions are used as synonyms. E-expertise bases on *information and communication technologies, as well as information analysis, decision-making, and artificial intelligence technologies.*[1] Today, e-expertise technologies are intensively applied to conduct public expertise, report expertise, and event expertise. The reasonability of using network technology represents a challenge for specialists in artificial intelligence (multi-agent systems, distributed decision-making systems—see dashed line in Fig. 1).

A subject (topic) of expertise concerns various properties of material and nonmaterial objects, events, phenomena, and processes in the past (retrospective expertise), present, or future (expert forecasting and strategic planning, prediction or even prophecy). Here are some examples illustrating the whole diversity of possible subjects (topics) of expertise: the areas of investments, a multi-stage technology of quality function deployment, the technical state of buildings and installations, the consumer properties of a product, a draft version of a legal act, a regional economic development strategy, the development prospects of an industry, the forecast of a research work or a research project, the opinion of a social group, the forecast of an election campaign, crisis forecasting, etc.

Generally, the topics and conditions of normative expertise are regulated rigorously. Contrariwise, *expertise* for making managerial decisions is initiated by a Principal or subjects of civil society.

In the context of e-expertise, the present book also operates on the following notions (see [67, 81, 88, 91, 101]).

---

[1] This work does not analyze networked expertise as a type of normative expertise, where an object represents a network (e.g., a computer network).

Polling of experts is interviewing them by an open or closed set of questions, e.g., in the form of questionnaires; special estimation scales and rating systems (semantic, fuzzy, graduated ones) can be assigned to each or some questions.

Trust is subject's hope that other subjects (individuals and/or organizations) interacting with this subject will come up to his expectations.

Report of an expert is a document (particularly, an electronic document) defining the progress and results of investigations conducted by him; the written contents of investigations and final conclusions regarding the issues explored by an expert; a type (source) of evidence.

Moderator (or *facilitator*) is a specialist performing several tasks: (1) coordination of collective expert procedures, (2) checking of agreements' execution by experts (in accordance with their user statuses), and (3) policing of expert information submission.

Public expertise (*in legislature*) represents certain procedures of verifying the compliance of draft laws and state authority decisions with existing standards and interests of society, rights and liberties of a man and citizen; these procedures are initiated and performed by state authorities or civil society institutions.

Networked (*electronic*) *strategic meeting* is an expert procedure mode under the chairmanship of a Principal, which serves for consentient elaboration of goals, formulation of problems, and choice of actions.

Networked (*electronic*) *expert community* is a group of legal entities and/or private experts representing different sectors of a society and rendering expert services; a networked expert community employs information and communication technologies that provide equal opportunities for all experts (regardless of their location) to participate in its activity.

Subject supervisor is an authorized representative of a Principal being responsible for problem statement and preparation of a generalized expert report.

Networked (*electronic*) *expert brainstorming* [58] is an expert procedure mode, which supports networked brainstorming under the supervision of a moderator. This mode is intended for rapid generation of nonstandard ideas and proposals.

Networks—Networks have been known from the earliest times. For instance, recall road networks in Ancient Rome, postal networks in the Middle Ages, railway networks, telegraph, or telephone networks. A new type of networks improved communication among people *ergo* promoted further progress. We give an example from the modern world. A global city can be comprehended as the concentration of intersecting "roads." Moscow pretends to the status of a business and financial center as the heart of communications, intersection of financial and other "roads."

As any phenomenon, the development of networks has been demonstrating positive and negative sides. Many scientists predict the forthcoming appearance of a new "informational (networked) society," with power seized by global networks and transnational corporations. They will control people, and everybody will have to fulfill their requirements. Moreover, such viewpoints have even yielded a new term, the so-called *netocracy*. It signifies a new form of society management, where the basic value consists in *information* and structures used to generate, store, process, and transmit information (rather than material resources such as currency, real estate, etc.). Besides information in its conventional interpretation, the process of society management essentially depends on various (public and nonpublic) institutions and electronic organizations, as well as on the collective unconscious. Actually, the last factor is revealed through spontaneous and purposeful mechanisms of e-expertise.

The intensive development of network institutions based on information and communication technologies highlights the problem of *intellectual property*. Any information allocated in the global network gets expropriated and becomes public. The related difficulties (in the first place, due to existing imperfections of the legal base) are clear. And expertise provides a relevant attribute in their solution.

Modeling of networks and conceptual expert models (hierarchical and cognitive ones) employ the well-developed apparatus of *graph theory*. Complicated objects and phenomena can be easily described by a set of elements and connections among them. A graph represents a set of nodes (elements) and a set of edges (connections, arcs) among them. Graph theory gives a convenient framework to model and cognize the structure of systems having different nature. Owing to visual methods, this framework seems intuitive and comprehensible for everybody (especially, for those having little to do with mathematics). In politics or economics, nodes correspond to factors characterizing a studied situation (inflation rate, product quality, competitive ability, public image, etc.), whereas arcs show the mutual influence of factors (strengthening, weakening, etc.). This approach was reflected in cognitive modeling methods [10]. Probably, among other formal conceptions, graph theory has mostly contributed to popularization of mathematics and application of mathematical methods in practice. Many researchers even treat graph theory as a universal tool of good communication between different scientific disciplines. On the other hand, in the recent decade, various network structures have been attracting the growing interest of theoreticians and practitioners (particularly, in the field of management).

Among network resources, the gradually ascending role belongs to *online networks* (social networks, expert networks, etc.) intended for communication support, opinions exchange, and data acquisition. However, they have recently become the objects and tools of informational control and the scene of informational contagion. In this context, we should mention the framework of graph theory and *game theory* as well.

Of course, online networks have certain benefits and shortcomings. The Internet was designed as a faster means of data transmission, but received further development mostly in the field of data acquisition. Nowadays, we browse the World Wide Web for necessary information (instead of visiting public libraries or bookstores). This seems very comfortable. On the other hand, a modern schoolchild, undergraduate or postgraduate student would hardly go to a public library to read or look into something important. Consequently, fewer people know about the things inaccessible through the Internet. Perhaps, in the foreseeable future, mankind will digitize all existing archives, books, and journals, and completely pass to electronic publications. But this is not the case today. The

Fig. 2 The interaction of
basic participants of "conventional" expert activity

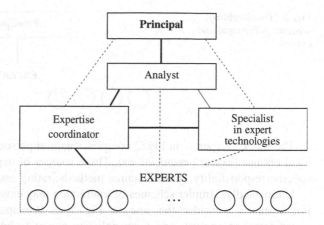

current generation of schoolchildren, undergraduate or postgraduate students lose touch with hard-copy materials and even the culture of their ancestors.

In the sense of information sources, an appreciable difference of any online social network from the Internet consists in the attitude of an individual to information provided by such network. Generally, the Internet provides anonymous information or information from well-known people (journalists, politicians, etc.). They are trusted owing to high rating or reputation gained. In a social network, a recommendation to watch some movie or purchase some product is made by "friends," i.e., people *trusted* not for their reputation or rating, but for personal relationships. In most cases, friends are not experts in products, yet enjoy more trust. In other words, a *social network* represents a source of personalized information (in contrast to Internet). Furthermore, a social network is a means of *communication*. Any individual appreciates the opinion of other people (including recognition, sympathy, and compassion). A social network is a means of support or even safety. Simply share your anxieties in a social network and receive "understanding."

The above-mentioned features apply to expert activity. Online expert networks serve for rapid acquisition and analysis of numerous opinions, organization of interaction among experts, generation of nonstandard decisions, and so on.

E-expertise (networked expertise)—The appearance of e-expertise is predetermined by three major factors: (1) accelerated changes in life conditions, (2) the development of group decision support systems, and (3) advances in information and communication technologies. Modern network information technologies (in the first place, Internet) organize communication of experts and an expertise coordinator. Moreover, they make a new method of expert data processing and collective expert decision-making.

Figure 2 illustrates the interaction of different participants within "conventional" expert activity.

Fig. 3 The simplified
scheme: A Principal and
experts

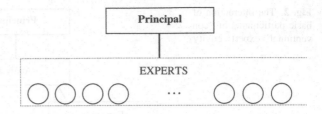

The block "Experts" in Fig. 2 comprises natural persons (individual experts) or legal entities (*expert organizations*). These sources of expert information employ specific responsibility, questionnaires methods, rating assignment methods, etc.

Undoubtedly, simpler schemes of interaction may exist in particular situations. For instance, consider the one in Fig. 3; a Principal acts as an expertise coordinator, an analyst, and a specialist in expert technologies. Such situations should be avoided, though.[2] E-expertise with very many participants requires the presence of an analyst to "translate" expertise results to a Principal.

Any participant of expert activity (a Principal, an analyst, a specialist in expert technologies, an expertise coordinator, a moderator, a subject supervisor, experts) can possess a complicated structure. As a customer and ultimate user of expertise results, a Principal may represent an individual or a collective authority; the institution of analysts may have a hierarchical structure; specialists in expert technologies may be natural persons or legal entities, etc.

Concerning fundamentally new capabilities, networked expertise excels the prevalent bureaucratic system of councilors (that "filter" and "dose" the incoming information of a Principal, see Fig. 4) in the following aspect. A Principal directly addresses any expert, a professional or expert community, a target public group, as well as easily organizes public expertise (see Fig. 5).

The novelty is not just that a Principal "breaches the front" of numerous advisors and accesses the "original information." E-expertise allows improving the quality of management by creating necessary prerequisites of stability and purposefulness in decision-making processes [88, 112]. In addition, expert procedures acquire new properties in principle. They guarantee appreciable reduction of risks in decision-making, require smaller costs under higher complexity, as well as accelerate formation of a "trust space" among experts. Figure 6 shows the scheme of an expert procedure, which incurs considerably greater costs without network usage.

At the same time, it seems incorrect to contrast "conventional" expertise with e-expertise. Actually, they are mutually supplementing and demonstrate some disadvantages and benefits (see below).

An example of comparing "conventional" expertise and e-expertise concerns the limited time and cognitive capabilities of a Principal.

---

[2] A possible exception is the following "expertise." A manager conducts a meeting with his deputies.

Fig. 4 "Common" expertise
in decision-making

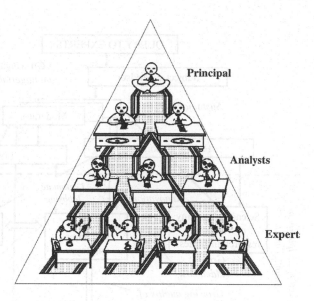

Fig. 5 E-expertise in decision-making

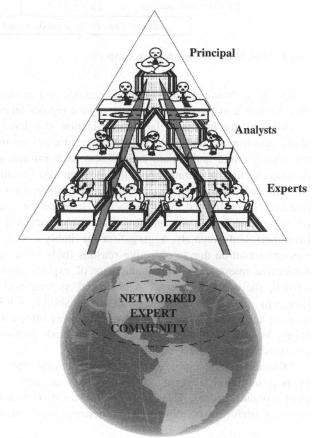

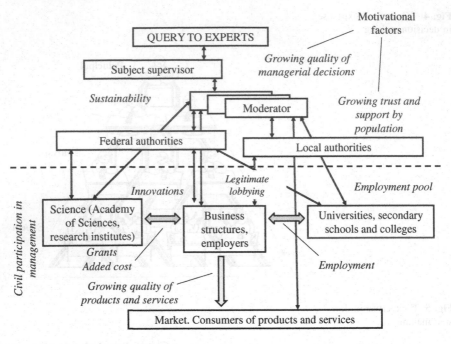

Fig. 6 Multiple factors considered in e-expertise

Ideally, a Principal should operate reliable and exhaustive information on any processes in a managed system (a state or a region, an enterprise, etc.).

However, the head of a large organization may have no detailed information about activities performed by an employee at a certain instant (more specifically, the former could and should not possess such information). Similarly, an army commander knows nothing about the location of a specific soldier during combat. And the need for *aggregation* (reduction) of *information* arises immediately. This function is performed by experts (as *information sources*) and analysts (as *converters of expert information* into a compact and transparent form for a Principal). Unfortunately, aggregation inevitably causes the risks of information misrepresentation due to objective reasons (bulky data, existing uncertainties) or subjective reasons (individual interests of experts and/or analysts). As is well known, the sustainability of a managed system goes down as management hierarchy grows. To improve such sustainability, a Principal must be able to acquire necessary (qualitative and quantitative) information about objects at any levels as soon as possible [88]. The stated task seems realizable only with the assistance of networked experts.

Different technologies of e-expertise provide the feasibility of addressing primary information (dialoging with experts, as well as expert, professional and public organizations). This improves the decision-making of a Principal in the sense of performance, objectivity, timeliness, and efficiency. Furthermore, such approach reduces the risks of misrepresenting the "real picture" by the corps of analysts and councilors of a Principal. This does not mean that a Principal would frequently utilize the original information (perhaps, he would not do it at all). But a Principal must have such feasibility in order to improve the sustainability of a managed system (using special expert procedures).

E-expertise possesses the following *advantages* (see [105] for the discussion of solutions generated by large groups):

- timeliness, accuracy, and the complete coverage of a problem;
- the representativeness of experts' opinions, as granting the most adequate reflection of interests pursued by intelligent and elite clubs, public, professional, and other groups;
- the feasibility of involving many experts such that each expert is a specialist in some subject domains (the feasibility of incomplete preferences' aggregation);
- the feasibility of direct communications and electronic collaboration [115] among experts;
- the development of management tools for nonfinancial motivation of experts (including estimation scales and rating systems);
- the presence of self-organization effects in an expert community, the appearance of "collective intelligence" (sometimes, researchers also adopt the terms of "local synergy of intelligence," "teleportation of ideas," and so on), including autonomous formation of experts' reputation;
- the feasibility of documenting and real-time processing of different communication aspects that reflect a sense of actions (connect-analysis);
- transparency, openness, responsibility, and democracy (see Sect. 1.3 of the book);
- the feasibility of activating special mechanisms of information processing (they implement the principle "Ten fools may generate a genuine idea!").

We have repeatedly mentioned that e-expertise incorporates potential obstacles to-be-overcome. In the first place, the matter concerns the loss of direct communication when experts meet virtually.

Second, the number of individual communications among experts may appear superfluous. A group of experts may comprise members with insufficient qualification or knowledge of a specific situation. Opinions of very many experts would hardly converge to a "common" aggregated assessment.

Third, there exists the problem of expert *finding*. This depends on the analysis method of expert data selected in given expertise (see Chap. 3). Another associated problem lies in rational grouping of experts (see Chap. 4).

And finally, analysts, coordinators, and specialists in expert technologies (of course, experts proper) can be interested in certain expertise results. The *problem of data manipulation* arises naturally (some subjects with individual goals exert a strategic influence on expertise results). The data manipulation problem is analyzed in Chap. 5.

The Structure of the Book

Chapter 1 ("E-Expertise") discusses the role of e-expertise in decision-making processes (Sect. 1.1). Next, the procedures of e-expertise are classified (Sect. 1.2), their benefits and shortcomings are identified (Sect. 1.3), and the efficiency conditions are considered (Sect. 1.4).

Chapter 2 ("Expert Technologies and Principles") provides a comprehensive overview of modern expert technologies. A special emphasis is placed on the specifics of e-expertise. Moreover, the authors study the feasibility and reasonability of employing well-known methods and approaches in e-expertise.

Chapter 3 ("E-Expertise: Organization and Technologies") describes some examples of up-to-date technologies to perform e-expertise.

Chapter 4 ("Trust Networks and Competence Networks") deals with the problems of expert finding and grouping by information and communication technologies.

Chapter 5 ("Active Expertise") treats the problem of expertise stability against any strategic manipulation by experts or coordinators having individual interests.

The Appendix provides supplementary information and some draft documents:

  I. Typical statute of networked expert community;
 II. Typical regulations of networked expert community formation within the structure/for the demands of federal authorities;
III. Assessment criteria for expert analysis organizations (think tanks);
 IV. Assessment procedure for expert ratings;
  V. Professional ethics code for networked expert communities;
 VI. Security problems.

The authors suggest several ways of perusal. A reader interested in the general conception of networked expertise can be confined to Chap. 1. To accumulate necessary information about networked expertise organization, one may terminate reading after Chap. 3. As a matter of fact, Chap. 2 represents a brief navigator over expert technologies. Finally, Chaps. 4 and 5 can be studied independently.

The authors are deeply grateful to A. Yu. Mazurov, Cand. Sci. (Phys.-Math.) for careful translation of this book into English, as well as for helpful remarks and feedback.

Chapter 1
E-Expertise

1.1 E-Expertise and Decision-Making

Traditionally, theoreticians represent the process of *decision-making* as a block diagram of choosing a certain decision from several alternatives to achieve formulated goals using accumulated experience and available resources. It is necessary to define assessment criteria for evaluating the alternatives with respect to the goals and a current situation. These criteria will assist in selection of a correct alternative. As a rule, such approach yields an inertial path, which is inefficient in some situations. A Principal often possesses exclusive information and has to make a single-alternative decision counter to the opinion of assistants and counselors. For instance, this happens in the following case. A Principal knows a specific situation in greater detail than the others and plans something ambitious disharmonizing with inertial ideas. When subordinates misunderstand a Principal, their motivation goes down and risks rise. To reduce risks and stimulate subordinates, a Principal should initiate a decision-making procedure such that his environment would "mature" independently to the level of his decision.

Unfortunately, many collectives lack for knowledge, experience, ideas or simply time for such maturing. In the described conditions, the lacunas of uncertainty can be filled by participation of a networked expert community in decision-making process and management process. This brings to a new paradigm of decision-making (see Figs. 1.1 and 1.2), where electronic *expert procedures* get involved in elaboration of decisions.

In the case of electronic expert procedures, the *management problem* admits the following formal statement. Find managerial decisions having the maximal *efficiency*, e.g., the ones improving the quality of governmental or corporate services under minimal costs, owing to the knowledge and insight of networked experts.

This calls for choosing an *optimal managerial decision* under a set of qualitative and quantitative factors affecting situation development including the factors that characterize the *feedback* from the customers of such services.

D. Gubanov et al., *E-Expertise: Modern Collective Intelligence*,
Studies in Computational Intelligence 558, DOI: 10.1007/978-3-319-06770-4_1,
© Springer International Publishing Switzerland 2014

Fig. 1.1 Involvement of electronic expert procedures in management processes

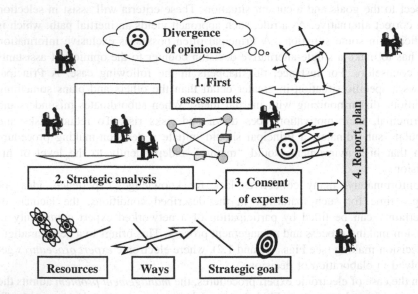

Fig. 1.2 Involvement of electronic expert procedures in collective decision-making

There may exist a hundred of critical factors (political importance, inflation rate, money supply, customer satisfaction, and computer literacy, to name a few). Furthermore, an efficiency criterion depends on goal factors and resource constraints (see Fig. 1.3) [79].

The block diagram of e-expertise (basic subjects, functions and their interaction) is demonstrated in Fig. 1.4.

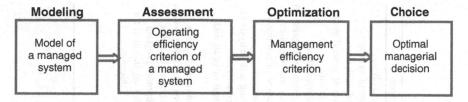

Fig. 1.3 The management problem: The general logic of statement and solution

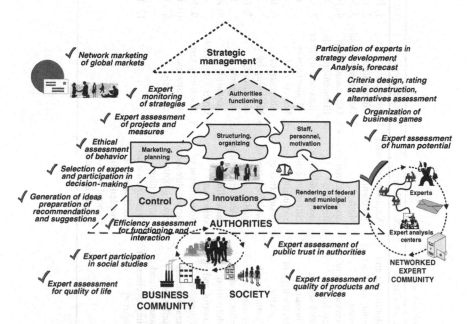

Fig. 1.4 The block diagram of e-expertise

Analysis of related problems and subsequent decision-making involving networked experts can be performed by different methods of systems analysis and strategic analysis [33, 73]. We combine some methods in Table 1.1.

E-expertise as a tool of making consentient collective decisions. Prior to discussing consentience aspects, we should focus on team work. A *team*, a collective, a department or a group of experts—all these terms describe a number of people with common business interests, being interconnected by a common view of goals and actions. The whole existing or still growing arsenal of methods—corporate culture development in different organizations, achievement of consent, improvement of employees' motivation, perfection of management—aims at business success of organizational structures whose elements are human beings.

A group of employees may be large or small, have network organization or not, be recently designed or possess well-established traditions. A group of employees may have an ordered or chaotic structure, be purposeful or uncoordinated, and dispose of

Table 1.1 Systems analysis and strategic analysis of problem solving

E.P. Golubkov	P. Drucker	D.A. Novikov	A.N. Raikov	N.P. Fedorenko	Yu.I. Chernyak
1. Problem statement	1. Purpose and expected results	1. Monitoring and analysis of actual state	1. Acquaintance with problem	1. Problem formulation	1. Problem analysis
2. Examination	2. Key elements to process design: time, resources, budget, major steps	2. Forecasting of evolution	2. Method specification	2. Definition of goals	2. Definition of system
3. Analysis	3. Roles and responsibilities of self-assessment team	3. Goal-setting	3. Strategic conversation in Situation center including	3. Data acquisition	3. Structural analysis
4. Preliminary judgment	4. Elements essential to success:	4. Choosing technology of activity	• construction of weighted tree of goals;	4. Elaboration of the maximal number of alternatives	4. Formation of general goal and criterion
5. Confirmation	• Utilizing an experienced facilitator;	5. Planning and resources allocation	• collective cognitive modeling;	5. Selection of alternatives	5. Goal decomposition, identification of demands in resources and processes
6. Final judgment	• Engaging dispersed leadership;	6. Motivation	• collective formulation of action directions; drawing optimal action plan	6. Modeling by equations, programs or scenarios	6. Identification of resources and processes
7. Implementation of chosen decision	• Encouraging constructive dissent;	7. Control and operative management	4. Organization and motivation of implementation	7. Costs estimation	7. Forecasting and analysis of future conditions
	• Using data to inform dialogue	8. Reflexion, analysis and improvement of activity	5. Control of implementation	8. Sensitivity tests (parametric analysis)	8. Assessment of goals and means
					9. Selection of alternatives
					10. Diagnosis of existing system
					11. Elaboration of complex development program
					12. Design of organization for goals' achievement

considerable or insignificant financial resources. Moreover, a detailed gradation system can be proposed for the sizes of a group; such system predetermines the difference in most efficient approaches and methods of e-expertise and coordination of decisions within this group. For instance, consider the following gradation system by the number of experts in a group: 2, 3, 4–12, 13–35, 36–250. Stating universal rules of successful team building seems impossible. The only universal rule claims that a group of people is always unique; thus, it requires individual management solutions with appropriate account of a problem situation. A new situation may call for application of a new combination of methods. This approach admits a simple explanation. The core of any organization consists in human beings; their nature is unique, and any community of human beings enjoys this property. The matter concerns managerial situations—the combinations of factors characterizing any problem situations always differ. An important feature of communities is the following. People are united by necessity, e.g., in order to guarantee their safety, to increase company's capitalization, to improve the quality of services, to form a certain idea, to solve a posed problem, or to realize some function.

A *group subject* is a collective of individuals with almost coinciding common interests, motivation, goals, ideas, and so on. *It is a team.* The interest of an individual in a team not necessarily obeys a collective interest (it can be hidden, latent, unconscious). On the other hand, the interest of a group subject does not represent the sum of individual interests; indeed, there is no point in grouping which gives nothing to individuals.

The activity improvement problem for teams possesses the interdisciplinary character. Philosophers and psychologists, economists and sociologists, physicians and mathematicians, managers and cyberneticians, doctors and biologists have been exploring it intensively. There exists a wide range of theoretical and practical tools of team building. Nevertheless, e-expertise mechanisms can serve for speeding up the degree of consentience among team members, as well as for accelerating team building.

The phenomena of interdependence, interpenetration and interaction in problem solving for different subjects (an individual, a group, an organization, etc.) and different types of consciousness (individual, group or mass consciousness) urge many researchers on accepting some general logical structure for preparation and making of managerial decisions. Such logical structure must be psychologically predetermined and substantially comprehended.

In this context, we also acknowledge the necessity of using computer-aided methods of elaboration, coordination and implementation of decisions (e.g., see [88, 91]).

Management science often puts the following question: "What is the optimal structuring of communication among different subjects in the sense of their most efficient coordinated interaction?" Such structuring allows analyzing the processes of informational interaction between employees and collectives in uniform terms and conceptions, with appropriate consideration of their specifics. This structuring brings to comprehension of the logical and psychological schemes of collective strategic subjects design. While considering the properties of strategic subjects, we

Table 1.2 The characteristics of collective strategic subjects

Aspects	Attributes	The characteristics of collective strategic subjects	
		Intrinsic	Polar
The teleological (goal-related) aspect	Type of goals	Original	Extrapolated
	The presence of primary goal	Yes	No
	The presence of common goals	The community of employees based on goals	Different goals of employees
	Separation of internal and external goals	Promotes identification of a group or problem	No separation
	Orientation	Social	Asocial
	Stability of goals	Regulated refinement	Spontaneous changes, randomness
	Coincidence between realizable and declared goals	Goal-adequate actions	Discord of goals and actions, variance of words and deeds
The functional (regulation and communication) aspect	The presence of regulation efficiency criteria (reflexion)	Criteria are specified by customers (feedback)	Criteria are specified by internal capabilities and motives
	Dominating needs	Spiritual, social	Material, economic
	Foundations of decision-making	Mental and ethical	Rational
	The process of problem solving	Solution of inverse problems	Solution of direct problems, dualism
	The leading ethical system	Combination of "good" yields "good"	Combination of "good" and "evil" yields "evil"
	Dominating component in the interaction of subjects	Collectivism	Individualism
	Settlement of conflicts	Consent, compromise	Aggression, stress
	Freedom of action	Rights and duties	Priority of rights
The structural aspect	Decomposition	The systems approach to control and management	Fragmentary structuring
The gnostic-analytical (cognitive) aspect	Search for the truth	Consent of interests and usability of deeds	Representationalism, search for adequacy
	Structuring of situations	The mutual influence of factors	Separation of factors
	Type of actions	Creator—from goals	Solver—from problems
	Source of actions	Intension, vision, intentions	External pressure, formal experience
	The methodological basis of collective decision-making	Cognitivism, convergence [88]	Situationalism, behaviourism

(continued)

Table 1.2 (continued)

Aspects	Attributes	The characteristics of collective strategic subjects	
		Intrinsic	Polar
The priority-problematic aspect	Formulation of problems The method of problem ranking	Conditioned by goals and factors Consentient and collective	Spontaneous, "common sense" Directive (dictated "from above")
The resource aspect (ways, means)	Choice of resources Connection between resources and business processes Choice of ways	Resources are allocated to priority problems Operation-wise connection Achievement of extraordinary, original goals Optimized "crooked" way	Resources are allocated for further development of "well-operating" systems Isolation of resources from goals and processes Realization of extrapolated, inertial ways

suggest the following primary aspects: the goal-related (teleological) aspect, the functional (regulation and communication) aspect, and the structural aspect. Furthermore, it is possible to examine additional aspects, *viz.*, the gnostic-analytical (cognitive) aspect, the priority-problematic aspect and the resource aspect (ways, means). The basic characteristics of administrative and corporate strategic subjects are combined in Table 1.2.

Strategic subjects are ideal samples, whose implementation would improve the efficiency of administration in the governmental and corporate sectors via creating the ambiance of leadership, trust, personal interest and motivation of employees in coordinated actions.

Here we emphasize a couple of aspects. On the one part, e-expertise can serve as a tool of consentient decision-making. On the other part, the ultimate goal not necessarily lies in reaching a common opinion of all experts (sometimes, this is simply impossible). Actually, the diversification and versatility of opinions form a positive property of any expertise. It provides a complex view of the subject of expertise, suggests unexpected alternatives, etc. The final decision still belongs to a Principal.

1.2 Classification of E-Expertise Procedures

There exist different bases to classify e-expertise procedures.

1. Individual and collective e-expertise. Generally, e-expertise is *collective*. Nevertheless, information and communication technologies (e.g., Internet, voice and messaging services like *Skype*) can serve for individual interaction between a Principal or an analyst and separate experts.

2. Passive and organized e-expertise. Actually, *passive e-expertise* consists in *expert monitoring* of print and electronic media and other network resources (social networks, databases, transferred information, etc.) using a set of criteria or indices. Passive e-expertise also includes simultaneous or subsequent analytical processing of such resources. Expert monitoring represents an indispensable attribute of complex decision support systems for managers at the level of a state, branches of industry, regions and large corporations. Monitoring can be permanent or problem-related. In the former case, the horizon of monitoring appears unbounded. Problem-related monitoring proceeds from the period of solving a specific problem. This typification of passive e-expertise seems rational due to different approaches to its preparation and implementation.

Organized e-expertise lies in a special-purpose query to a set of experts. It is initiated by a Principal, a group of his analysts or another professional subject (for instance, organizations focused on public opinion analysis, media study, and so on).

3. Autonomous and non-autonomous e-expertise. Within the framework of *autonomous e-expertise*, experts adopt fixed procedures and information and communication technologies to form a certain collective opinion independently.

The result of non-autonomous e-expertise represents a set of opinions processed by analysts or provided directly to a Principal (if the set is sufficiently small).

A primary advantage of many types of e-expertise concerns autonomy; therefore, we discuss this property in detail. A common situation in practice is when experts fail to reach a consensus. The coordinator of e-expertise and/or a Principal have to analyze the whole variety of their (possibly, opposite) opinions. However, the existing variations in expert assessments are not necessarily inauspicious. The fact of missed consensus may give a fruitful ground for a Principal. Expert procedures stimulating generation of various ideas and opinions are called *divergent*. On the other part, *convergent* expert procedures call for a common opinion.

An important way to reach a consensus of experts is utilization of corresponding moderation procedures, provision of structural conditions guaranteeing stable convergence of an expert procedure to a consensus [88]. One approach to reach a consensus concerns "shifting the burden" of problem solving to experts, with subsequent receipt of a common opinion from them. And this opinion is then used in decision-making. We call such e-expertise mechanisms *autonomous*. What are the conditions stimulating autonomous operation of experts with suggestion of the best decision to a Principal?

Suppose that experts have to generate the best decision in a specific situation. Due to different education, experience, etc., some experts display higher degrees of competence in one subject domain, others do so in another subject domain. Everything depends on a conjectural situation of decision-making (the set of feasible situations).

A Principal prefers that, in any situation, the decision elaborated by experts demonstrates the highest efficiency. In other words, it is strongly desired to maximize the efficiency of a group of experts for each situation.

Let each expert know his efficiency but have no idea of the efficiency levels of other experts. And so, an expert can misrepresent information. In addition, assume that all experts identify a current situation adequately. What are the recipes to stimulate experts to suggest the most efficient decision in any situation? Consider the following mechanism. In a current situation, experts receive a simple offer: "Each of you reports to the rest experts your decision and the corresponding efficiency in the current situation. (Recall that each expert knows the actual efficiency of certain decisions suggested by him in a specific situation). Next, you report the decision ensuring the maximal efficiency in the current situation, and a Principal gives incentives to all experts proportionally to the efficiency level of the suggested solution."

Really, the stated mechanism is simple—experts choose themselves which decision to suggest, i.e., they work autonomously. It was shown [26] that, in an autonomous mechanism, truth-telling appears beneficial to experts!

Concerning the advantages of autonomous expertise mechanisms, we should mention (1) the "relief" of an analyst and/or Principal (owing to direct receipt of an optimal decision) and (2) strategy-proofness (see Chap. 5).

4. Expertise with a fixed or arbitrary number of experts, with a fixed or arbitrary staff of experts. Consider four cases: *the number and staff of experts can be fixed or arbitrary (without regulations)*, see Fig. 1.5.

Case 1 (the number and staff of experts are fixed) generally corresponds to "conventional" expertise. Next, case 2 (the number of experts is fixed, whereas their staff appears arbitrary) agrees with direct addressing an electronic expert community with freedom in the choice of opinions.

Case 3 (the staff of experts is fixed, without regulation of their number) corresponds to random selection of a certain number of experts from professionals in dedicated fields.

And finally, case 4 (the number and staff of experts are arbitrary) matches direct addressing an electronic expert community with subsequent analysis of all opinions.[1] Such expertise may involve citizens not included in any registers of experts. The matter concerns public (social) expertise, mob and crowd wisdom or crowdsourcing. For instance, we mention public feedback on some draft law or the traffic jamming problem in a megalopolis. In this case, it is possible to accumulate several thousands of offers and proposals within a small period. However, such proposals and opinions require adequate ordering, e.g., using e-expertise mechanisms with assistance of employees and experts in the specifics of managerial activity (we mean managers responsible for final decision-making).

5. Expertise with regular or random experts. No doubt, this classification base for e-expertise procedures strongly correlates with the previous one. Experts can be professional or random. In the former case, expertise grounds on reliable, authoritative and certified experts verified through appropriate registers of experts. The latter case is widespread in e-expertise, and greatest difficulties arise during analysis of professionalism and competence of experts. However, the issue of throwing a "random opinion" in expertise by a "random expert" becomes topical in certain techniques of decision-making. A "random opinion" may play a major role in electronic brainstorming with networked experts [58], when it is necessary to incorporate information from another problem domain into expertise.

The set of classification bases for e-expertise procedures can be extended. For instance, we can classify procedures by the formal *responsibility of experts*, the presence or absence of *financial incentives* of experts, etc. Expert procedures also differ in the methods of experts' selection and grouping, experts polling, as well as in the methods of acquisition and processing of expert information (see Sect. 2.2).

---

[1] We underline an important aspect—increasing the number of experts may impair expertise. Actually, everything depends on some properties of expertise (in particular, its complexity, multi-aspect character, and subject). Assessment of "simple" objects can employ very many experts of almost any qualification levels (see examples in [105]). Yet, in "complicated" situations, one should invite several professionals.

Fig. 1.5 The set of experts

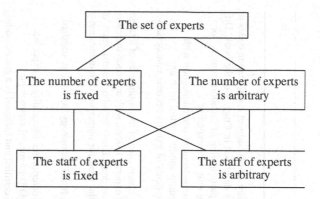

1.3 E-Expertise: Capabilities, Restrictions and Conditions of Applicability

In a specific situation, *conditions of applicability* are defined by balancing the benefits and shortcomings, the capabilities and restrictions of e-expertise.

We have emphasized that, in this book, e-expertise is comprehended as a generalization, development and expansion of "conventional" expertise owing to modern information and communication technologies, primarily artificial intelligence technologies and network technologies such as Internet. Thus and so, the comparative analysis of conventional expertise and e-expertise (see Table 1.3) mostly bases on the capabilities of network information and communication technologies (with their inherent advantages and drawbacks). Notably, we embrace the reduction of "world's diameter," the appearance of "collective intelligence," group intelligence or even group and crowd wisdom, the capabilities of online communication with almost unbounded number of subjects and global network search, etc.

Once again, we underline the relevance of the following aspect. Generally, experts participating in e-expertise express their opinions in a free form. This requires special methods of opinion processing. Accordingly, e-expertise is close to monitoring of electronic and print media, network resources (and the difference consists in the purposeful character of e-expertise). In both fields, one has to apply semantic analysis technologies for natural language texts. At the same time, it is possible to use e-expertise modes with linguistic scales, questionnaires, and so on. Such tricks appreciably reduce the labor intensiveness of analytical processing of expert information (see Chap. 3).

1.4 Efficiency Conditions of E-Expertise

We outline basic groups of conditions for the applicability and efficiency of e-expertise. Actually, the substantive *conditions of* its *applicability* have been discussed in Sect. 1.3.

Table 1.3 Conventional expertise versus e-expertise

Characteristic	Conventional expertise	E-expertise	The properties of e-expertise
From the viewpoint of a Principal			
The costs of expert finding	High	Low	Sometimes, it suffices to search by key words, subject classification codes or (generally) to apply technologies described in Chap. 4 of the book. Different techniques may stipulate participation of "random experts" or crowdsourcing, which is feasible only in e-expertise
The costs of expert group organization ("assembly" of experts)	High	Low	There exist no geographical constraints—remote communication with experts is possible. Expert procedure admits control from any geographical point. Spatial movements of experts become unnecessary, leading to cost saving
The costs of expertise conduct	High	Low	These costs can be high under strict observance of the boundaries of a target group or reaching a consensus of experts. Indeed, such approach increases the science intensiveness of expert procedures, the labor intensiveness of analytical data processing, and the need for numerous communication iterations
The period of expertise organization and conduct	Large	Small	Provided that e-expertise has a fine-tuned technology, as well as standard goals and subject (topic)
The capabilities of large-scale involvement in expertise	Bounded	Almost unbounded	Sometimes, the quantity of experts deteriorates the quality of expertise. Quality improvement calls for skills in special technologies
The capabilities of target group representativeness	Medium	High	Provided that networked experts are certified and included in a corresponding register of experts
The capabilities of typical (cut-and-dried) decisions	Medium	High	In many cases, the "issues" of e-expertise complement the issues of conventional expertise. E-expertise requires more accurate regulation and standardization
The capabilities of individual verification and expert qualification assessment	Medium	High	E-expertise by information and communication technologies enables detailed documentation of decision-making process. Thus, expert's work and intellectual property are identified with higher accuracy, while expert's status is assessed using content analysis tools

(continued)

Table 1.3 (continued)

Characteristic	Conventional expertise	E-expertise	The properties of e-expertise
The capabilities of expertise moderation (control)	High	Medium	A networked expert procedure implies "rigid" moderation as a prerequisite of a success
The dynamics of interaction with experts	High	Medium	Well-timed data acquisition and processing
Expert opinions' personification	High	High	E-expertise by information and communication technologies and special techniques enables detailed documentation of expertise process. Thus, expert's work and intellectual property are identified with higher accuracy, while expert's motivation is increased
The influence of experts conformism on results	High	Medium	Networked experts are spatially distributed. This feature prevents from conformist behavior and facilitates original judgments
The necessity of careful methodological preparations	High	Very high	Generally, conventional expertise fixes some representation form for the opinions of experts (a value, a feasible range, the number of answers to a closed question, etc.).Such unification seems difficult in e-expertise
The labor intensiveness of expertise results processing	High	Low	The total costs of e-expertise can be appreciably high in the case of designing special methods of expert data processing, training of experts and moderators
From the viewpoint of an expert			
The starting costs of participation (access price)	High	Low	For conventional expertise, the membership in an expert community is required. For e-expertise, the prerequisites are Internet access and inclusion in a register of experts
Democracy	Low	High	Everybody can "vote," and any opinion can be received and fixed
The costs of expert's status (reputation) maintenance	Low	High	It takes years for gaining reputation (in business sphere, politics). Yet, it can be easily shuttered. An expert should always make himself known. This task seems simpler for a networked expert
Technological level	Low	High	A certified networked expert possesses a free access to unique computational resources, expensive simulation technologies, and hyper efficient analytical services
Awareness of experts	Low	High	A networked expert possesses online access to exclusive up-to-date information

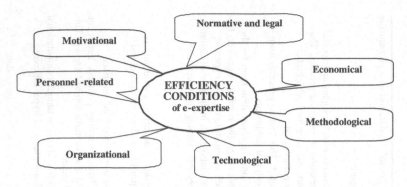

Fig. 1.6 Efficiency conditions of e-expertise

Let us identify the following *principles of expertise* (including e-expertise), see Sect. 2.3, and the following *efficiency conditions* (see Fig. 1.6):

(1) *Normative and legal conditions*—the feasibility in principle to organize e-expertise; the well-defined status of an expert and mandatory consideration of an expert report in decision-making; the regulated responsibility of the key participants of expert activity (an expert, an analyst, a Principal)–see Sect. 3.9;

(2) *Economic conditions*—the presence of a budgetary system, which guarantees financial control and cost saving in e-expertise. The coordinator of expertise has to build into the budget financial resources for expert activity organization (including incentives for experts, analysts and expertise organizers, see Sects. 3.10 and 3.11). The budgetary system of e-expertise must be result-oriented;

(3) *Methodological conditions*—the presence of a certain complex of e-expertise technologies, specific models of an assessed object, efficient methods of expert finding and accelerated reaching a consensus of experts, as well as techniques for creating structural conditions which guarantee the convergence of expert procedures to given targets. In addition, methodological conditions stipulate for certain methods of expert information acquisition and processing, conceptual modeling, optimal decision elaboration and presentation of an expert report to a Principal;

(4) *Technological conditions*—the presence of some infrastructure and direct access to appropriate information and communication technologies including intelligent information technology. A major condition here concerns operation of a software portal (as a cloud computing system) for managing a register of experts with rating systems, maintaining communication with experts, constructing analytical models for substantiation of draft decisions;

(5) *Organizational conditions*—the correspondence between the organizational structure of e-expertise coordinator and its technologies; such structure must include, at least, institutions of subject supervisors, moderators and experts (expert communities);

(6) *Personnel-related conditions*—the presence of personnel being able to provide methodological conditions (including analysts, cognitivists and knowledge engineers that convert expert information to a convenient form for Principal's comprehension, numerical simulation and modeling).

(7) *Motivational conditions*—in addition to incentive schemes for experts, it is necessary to construct an appropriate management system for nonfinancial motivation of experts; such system bases on expert ratings, reporting exclusive information to experts, involvement of experts in decision-making processes, etc.

Nowadays, the majority of countries face serious problems in the field of normative and legal conditions of e-expertise and expertise in general. Still, the status of expertise in state authorities has not been completely defined. Is it mandatory to consider expertise results during project implementation and elaboration of managerial decisions?

While analyzing the efficiency conditions of e-expertise in a specific case, one should identify (a) *internal problems* that can be solved within an organization (the customer and/or coordinator of expertise), i.e., independent creation of necessary conditions is possible, and (b) *external problems* to-be-treated, e.g., at the level of federal legislature.

Chapter 2
Expert Technologies and Principles

This chapter pretends to be a navigator over *expert technologies*. It describes the basic stages and methods of expertise, methods of expert grouping, typical errors, as well as the general technology of expertise organization and its principles. Finally, we discuss some prediction problems. The exposition *par excellence* proceeds from generalization of well-known classical statements (see *the theory of expert appraisals*) with emphasizing the specifics of e-expertise.

2.1 Stages and Methods of Expertise

Expertise [52] comprises the following *stages*:

(1) Deciding on the necessity of expertise and defining its goals by a Principal;
(2) Selecting and appointing the staff of a working group (WG) by the Principal. Generally, a WG consists of a scientific supervisor and a secretary; in the case of e-expertise, it also includes a subject supervisor, moderators (facilitators) and specialists in expert technologies);
(3) Elaborating requirements specification for expertise and approving this document by the Principal;
(4) Working out a detailed scenario (i.e., a procedure) of acquiring and analyzing expert opinions (comments, assessments), possibly, with modeling;
(5) Selecting experts;
(6) Forming of an expert commission (group);
(7) Acquiring expert information;
(8) Performing computer analysis of expert information (in several rounds according to an expert procedure scenario—repeating this stage and the previous stage);
(9) Performing final analysis of expert opinions, with interpretation of the results obtained and preparation of an expert report for the Principal;
(10) Official closure of the WG including expert report approval by the Principal.

D. Gubanov et al., *E-Expertise: Modern Collective Intelligence*,
Studies in Computational Intelligence 558, DOI: 10.1007/978-3-319-06770-4_2,
© Springer International Publishing Switzerland 2014

The specifics of e-expertise must be accounted at Stages 4–7. Common problems of expert finding and grouping are explored in Chap. 4. Next, the peculiarities of networked acquisition of expert information are considered in Chap. 1. For the rest stages (1–3 and 8–10), one may and should adapt well-known results from the theory of expert appraisals.

We also suggest the following typical scheme of expertise, see Fig. 2.1. Within the framework of this approach, the specifics of e-expertise must be considered at Stages 1, 4, 5, 6, 7 and 8.

As a rule, e-expertise represents collective expertise. Researchers separate out the following *features of collective expertise*:

- guaranteeing the maximum possible apprehension of a situation;
- revealing sure uncompetitive decisions;
- revealing true "theoretical" judgments and hypotheses;
- obtaining objectified assessments with weighty evidence;
- obtaining experts appraisals of higher reliability.

When collective expertise has electronic (networked) implementation, the listed features are supplemented by methodological and technological tricks described in Chaps. 1 and 3 of the book.

One can propose the following *quality assurance conditions for expert information*:

(1) the presence of subject supervisors enjoying the trust and understanding of their Principals;
(2) the presence of methodologists mastering the theory and practice of decision-making support in uncertain conditions;
(3) the presence of an expert commission (group) with professional knowledge of the subject (topic) of expertise and extensive practice of expert work;
(4) the presence of an analytical group with high-level skills of expertise organization and conduct, cognitive modeling and quality management methods, acquisition and processing methods for expert information;
(5) reliable expert information extraction;
(6) correct treatment and analysis of expert information using conceptual computer simulation.
    Perhaps, we should append an important condition (*the principle of integrity*):
(7) complete and holistic coverage for the properties of the topic of expertise (assessed object) by professional competencies of experts (with feasibility of involving experts from allied fields) [90].

Consider e-expertise performed by a fixed collective of professional experts (we refer to the classification of e-expertise procedures in Sect. 1.2). A distinguishing characteristic of such expertise (against "conventional" expertise) lies in wide usage of information and communication technologies for expertise. Imagine that an expert audience is not a priori fixed. In this case, ensuring the desired professional level of experts, reliability of expert information and complete coverage

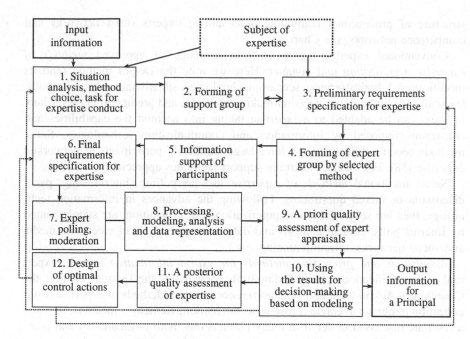

Fig. 2.1 A typical scheme of expertise

of the topic of expertise require additional analysis and efforts from coordinators of
e-expertise.

Organizational *methods of expert selection and forming of expert groups*,
namely,

- assignment methods;
- mutual recommendation methods ("snowball" methods);
- sequential recommendation methods;
- nomination methods for professional units and organizations by collectives;
- documentation methods;
- testing methods, etc.

have been intensively studied in scientific literature. E-expertise can employ traditional methods or other methods dictated by its specifics. For instance, we
mention the *"self-nomination" method* when expertise engages volunteers or all
persons concerned (particularly, in public expertise). Another example is the
"imperative assignment" method, i.e., inviting experts from allied fields with high
incentives and motivation.

Certain methods of expert selection get predetermined by the goals of expertise
and its subject with proper consideration of intended methods of expert information processing, and so on. Furthermore, one should not ignore the established

structure of professional communications among experts (trust networks and competence networks, see Chap. 4).

Conventional expertise possesses a wide range of approved *methods of expertise organization and conduct*. Here we note the Delphi method (and its modifications), brainstorming, scenario technologies, situational analysis methods, trees of goals and criteria, decision-making matrices, and prediction graphs. Many of them can be adapted to e-expertise taking into account the capabilities and constraints imposed by information and communication technologies. Some methods become difficult to use for a large number of participants of networked expertise [58]. For others, electronic support facilitates appreciably.

Select traditional *methods of expertise* such as polling, interviewing, panel discussion or mixed questioning. Following the advances in information technology, they are subjected to computerization and, in addition, are supplemented by Internet polls, content analysis and data mining, analysis of electronic media and social networks, teleconferencing, etc.

Acquisition and processing methods for expert information include expert rankings, classifications, paired and multiple comparison, the Terstone method, the Churchman–Ackoff technique, and preference vectors. Methods of expertise result definition (embrace averaging, row-wise summation, Kemeny's medians, to name a few) are discussed in detail in many publications, both in theoretical and practical aspects. Of course, their choice is restricted by specific methods of expert information acquisition and processing. All common methods can serve for processing of e-expertise results (depending on its form). A quite another pair of shoes is that e-expertise may yield results in an arbitrary form (e.g., texts, see Sect. 1.3). Extraction of well-structured "quantitative" information requires analytical work with certain methods of intelligent data analysis [39, 40], semantic analysis of knowledge mining from natural language texts.

There exist four basic types of *procedures intended for expert information acquisition*:

(1) one-shot procedures with direct getting of experts' comments and appraisals;
(2) repeated procedures without direct interaction among experts (e.g., repeated acquisition of comments and assessments from experts, mass polling of experts, current expert monitoring);
(3) iterative procedures with direct interaction of experts (e.g., strategic expert conversations [88] or panel discussions in conventional expertise, electronic brainstorming [58], discussion forums or blog posts in the case of e-expertise);
(4) iterative procedures without direct interaction among experts (Delphi procedures).

Application of such procedures in e-expertise proceeds from modern information and communication technologies, as well as information analysis and collective intelligence technologies, see Chap. 3.

The existing theory of expert appraisals provides fruitful results in many fields, namely,

- analysis of expertise results including consentience study of expert appraisals or estimation of the consistency index of experts;
- assessment methods for experts (rating systems, a priori and a posteriori assessment);
- solution of different problems such as inaccuracy and inconsistency, reliability and fuzziness of expert appraisals, missed consentience of expert appraisals during collective expertise.

All of them can be applied in e-expertise.

Expert activity is accompanied by the following typical *errors*:

- incorrect choice of a method for conducting expertise and further processing of expertise results in a specific situation;
- depreciated role of the normative, moderative and motivational provision of expertise;
- overstated capabilities of expert methods including the potential of network technologies;
- superfluous attention on "common sense" (far from always, such approach generates sensible ideas);
- usage of incompetent experts or incorrectly selected experts;
- ambiguous statement of the problem for experts;
- rejection of multi-alternative expertise, closeness of expertise;
- superfluous attention on quantitative estimates and formal models, to the prejudice of qualitative estimates and conceptual models;
- violated principles of measurement theory (incorrect operations with expert appraisals), incorrect analytic treatment of expertise results;
- insufficient or superfluous informational interaction of experts, deficient usage of methods and tools of virtual collaboration [115];
- disregarded strategic behavior of experts (including conformism—see Chap. 5 of the book);
- incorrect interpretation of expertise results (sometimes, with substitution of expertise for decision-making process, etc.).

2.2 General Technology

Clearly, the system character of expertise is realized via complex choice of different elements in the technology, motivation and moderation of expert opinions with proper consideration of a specific problem situation. In what follows, we describe nine sets of elements; expertise organization requires selecting certain elements from these sets, which possess required properties [101].

1. Formation methods of expert groups. There exist numerous techniques to compose expert groups including assignment, mutual recommendations or the so-called snowball. At the same time, e-expertise calls for tighter regulations and a wider range of expert finding methods (for a detailed discussion, we refer to Chaps. 3 and 4).
2. Candidates' selection criteria for expert groups. It is possible to use various criteria such as creativeness and activity of an expert, the number of publications, the h-index, the number of approved expert forecasts, the level of competence in a corresponding field, practical experience, ability to perform decomposition and synthesis, stability of opinions. The specifics of e-expertise are reflected in Chap. 4.
3. Means and methods of expertise (e.g., commenting, strategic conversation, divergent brainstorming, polling with linguistic scales, interviewing, depth interview, and so on).
4. Elementary types of expert appraisals including verbal judgments, numerical score, interval score, (single or multiple) point estimates, grouping, paired comparison, multiple comparison, ranking, preference vectors, functional estimates, etc.
5. The basic principles of expertise conduct. In the first place, expertise publicity or the independence of experts from other subjects during their expert activity, multiple-alternative expertise (see Sect. 2.3).
6. Algorithmic operations and procedures for indirect generation of a selected type of expert appraisals. For instance, the procedure of paired comparison and hierarchy analysis [99], cognitive models and SWOT-analysis, matrix comparison, the Churchman–Ackoff technique, the von Neumann–Morgenstern method, classification and multiple comparison, and others.
7. Elementary organizational procedures and methods for the activity of expert groups to accumulate new knowledge from experts (e.g., brainstorming, focus-group, problem-related group, the Delphi method).
8. Choice and analysis methods for error indices to perform a posterior quality assessment of an expert report.

While choosing certain elements of the general technology of expertise in a specific case, one should have in mind the following:

(1) the specifics of an object, event, a material or process under consideration;
(2) the level of problem comprehension by an ultimate user (a subject supervisor, an analyst and/or a Principal) and an expert;
(3) the goals and tasks of expert activity participants, as well as a series of constraints:
   (3.1) financial constraints (payment for all expertise work, individual honoraria of independent experts);
   (3.2) temporal constraints (forecast period, elaboration period of an expert report);

(3.3) personnel-related constraints (the capabilities of expert selection and expertise coordinator finding);

(3.4) normative and legal constraints (the status of an expert report: mandatory or recommendatory);

(3.5) motivational constraints (first of all, nonfinancial incentives).

9. Methods of accelerated reaching a consensus among the participants of an expert procedure regarding their goals and ways of actions, forecasts, opinions, conclusions, as well as most efficient managerial decisions. Different structuring techniques of expert information, which base on some methods of thermodynamics, solution procedures for inverse problems on nonmetric spaces or even methods of quantum mechanics [62, 88, 90], may provide appreciable assistance here.

2.3 Principles of Expertise

Following [67, 81, 101], consider fundamental principles for any modern type of expertise including e-expertise (specific difficulties of e-expertise arise immediately as one endeavors to meet the requirements below).

1. **Requisite variety**. All subjects participating in elaboration of an expert report must have the opportunity of choosing any elements of expertise technology (including selection of expertise methods, types of expert appraisals, ways of expert finding and polling, as well as accuracy indices). Furthermore, this principle guarantees the freedom of actions for all participants of expert activity based on an appropriate legal base). On the other hand, for the sake of expert procedure convergence and accelerated reaching a consensus, all experts must follow a uniform methodology or a system of interconnected methodologies of multistage expert processes and recommendations of moderators.

2. **Publicity of expertise** (the stages of elaborating requirements specification for expertise, discussion of final results and decision-making) implies publishing of different materials (except confidential or overhead information stipulated by legislation and normative acts). They include

   • the register of experts, members of expert commissions and local or federal authorities;
   • documents regulating organization and operation of basic participants of expert activity;
   • the conditions of tournaments and auctions, as well as the rules of query submission;
   • the list of conducted expertise procedures;
   • the results and materials of expertise (in the case of socially-oriented expertise).

3. **The system character of expertise and its technology** lies in that, as a type of activity, expertise represents an element of general decision-making. This principle manifests itself in definition of boundaries for the subject (topic) of expertise, in precise substantiation of the tasks and goals of expertise, as well as in specification and consideration of external links in an assessed object. It is necessary to balance the order and chaos in informational processes accompanying decision-making [112].

4. **Quality control for expert appraisals** (the need for feedback in expertise). The coordinators of any expertise must contrast the appraisals of different experts with each other and with the reality. Quality control plays a crucial role in experts' rating and their selection for further expertise procedures.

5. **The system character and continuity in expertise conduct**:

   - the systematic consideration of expert messages and selection of experts from potential candidates for expertise conduct based on the principle of quality control (feedback in expertise);
   - continuous improvement of the methodological, informational and organizational support of expertise (update and perfection of databases, background and normative materialsù as well as information on the staff and qualification levels of expert communities);
   - acquisition and analysis of information on the consequences of decisions made on the basis of conducted expertise;
   - spot checks of expertise quality, implementability assessment for previous forecasts of experts.

6. **Independence of experts from other participants of expert activity** is achieved via

   - appropriate normative and legal provision (adoption of rights, duties and responsibilities for participants of an expert activity);
   - professionalism and high mental and ethical qualities of experts;
   - involvement of experts without individual interests in certain results of expertise;
   - formulation of definite rules of expert selection and exclusion from expert commissions;
   - formation of certain mechanisms neutralizing and/or compensating external factors with one-way effect on experts' opinions;
   - strategy-proofness of expert procedures (see Chap. 5 of the book).

The independence within established authorities must be maintained by current civil legislation and other normative acts which provide for punishment for any pressure on an expert or interference in the activity of an expert or expert commissions.

7. **Legal balance** concerns the parity of rights, duties and responsibilities of each expert activity participant within legal boundaries.

8. **Objectivity or eliminating "the conflict of interests"** among participants of expert activity. The following rules fix contraindications to involvement of specific subjects in independent expertise:

   - experts do not assess objects whose representatives have well-established relations with them treated as community/conflict of interests;
   - representatives of an assessed object do not participate in its expertise as experts or coordinators;
   - representatives of expertise customer do not participate in settlement of issues, where they have individual interests;
   - the number of staffers in an expert commission (here a staffer means a representative of an organization maintaining operation of such expert commission or a representative of a subordinate organization) does not predetermine decisions for the benefit of this organization.

9. **Personification of experts.** During expertise, the status of an expert as a high-level specialist in an appropriate field comes before his belonging to a certain organization or subordination to a Principal. In the case of e-expertise, this principle often holds true.

10. **Single-shot expertise.** Actually, repeated expertise of a same object is allowed in the following situations:

   - by a decision of superior authorities (for expertise customer);
   - by a legal decision;
   - if decision-making is impossible due to uncertain results of previous expertise.

   There must be a clear provision for other cases of repeated expertise in the normative document of expertise. Moreover, repeated expertise is allowed only with another group of experts; the materials of the previous expertise are considered only at the stage of decision-making.

11. **Confidentiality of expertise.** Without permission of interested subjects,[1] the customer and coordinator of expertise must not announce (a) an expert making a certain appraisal (to representatives of the object of expertise) and (b) the authors or source organization of specific materials submitted to expertise (to experts). These representatives must have no influence on the motivation or financial stimulation of experts.

12. **Democracy of expertise.** Formation of temporal or permanent expert commissions requires

   - conducting an open tournament of candidate experts (any exceptions must be mentioned);
   - updating the staff of permanent expert commissions based on expert ratings.

---

[1] In the case of e-expertise, such permission can be "implemented" by default. For instance, when an expert shares his opinion during open online debates, he provides public access to such opinion.

13. **Responsibility of expert activity participants and their legal safety**. It is necessary to make a clear provision for:

- the responsibility of an expert for his messages and usage of confidential information accessed during expertise;
- the responsibility of customer for ignoring expertise results in decision-making (in the case of material damage, financial losses, etc.).

Such responsibility is maintained as follows. A decision of an expert or expert commission causing material or moral damage, financial losses can be the subject of a legal action and further reimbursement of damages.

Legal safety of all expert activity participants is guaranteed by legislation and corresponding realization mechanisms, i.e., local normative acts.

2.4 Expert Forecasting

At all times, people strive for reducing the impact of uncontrolled factors on the results of their activity (by acquiring additional information on the unknown or incompletely known). Perhaps, this aspect explains the popularity of various forecasts (weather forecasts, marketing forecasts, economic forecasts, scientific and technical forecast, etc.). According to Merriam Webster Dictionary, a forecast is a prophecy, estimate, or prediction of a future happening or condition.

There exist several groups of *forecasting methods* for practical application [101]. For example, these are extrapolation methods, strategic planning methods, expert appraisal methods, logical simulation methods.

Extrapolation methods concern analyzing major tendencies in certain developmental aspects of a society, science and technology, forms of labor organization, industrial engineering, etc. Various information on the history and further development of phenomena and processes is studied, compared and transformed in numerical form. Subsequently, certain regularities and laws are extended to future periods (extrapolated). The corresponding conclusions serve as the foundation of a resulting forecast (generally, the evolution of considered objects).

Strategic planning methods. A directive sets a required future state of an object. For instance, during a strategic discussion, participants are asked about the future level of a company, industry, department, etc. Using their "forecast," a strategic plan of directions and measures is compiled then.

Expert appraisal methods. Essential information for forecasting bases on the opinions of highly-skilled experts in dedicated fields. Such opinions are formulated independently and accumulated by specialists. Next stage lies in their statistical treatment and strategic analysis. As a result, one obtains a snapshot of the future state, as well as possible scenarios.

In other words, *expert forecasting* can be treated as a forecasting method and as a type of expertise. Therefore, e-expertise may serve for forecasting. Unfortunately, networked technologies are still not intensively adopted in forecasting problems, although a series of research works demonstrate the efficiency of collective intelligence, mob or group wisdom, etc.

Logical simulation methods imply designing logical models that draw analogies between heterogeneous phenomena or processes, as well as generalize data on scientific, technological, economic or social development.

Researchers distinguish between the descriptive approach and the normative approach to forecasting [77]. The *descriptive* approach defines possible future states of a forecasted object. An example is a forecast of energy development (the appearance of new energy sources, the usage of existing energy sources after several years).

A problem of *normative forecasts* consists in choosing the ways and periods of reaching desired states of an explored object in future. A normative forecast represents prophecies attracting interest and stimulating some actions. For instance, imagine that we have a normative forecast of energy development. Then it is possible to pose the forecasting problem for the energy sector of a country. Here the ultimate aim consists in guaranteeing a required level of per capita energy consumption under certain constraints on available nonrenewable resources.

There are two "extremes" in the impact of a forecast on the pace of developments. A *self-implementing forecast* is a forecast which becomes reliable only by having been made. If we predict a rise in inflation due to uncontrolled growth of money supply, this rise occurs *per se*. A self-canceling forecast is a forecast which becomes unreliable (or avoidable) only by having been made. In the middle of the 1980s, Academician N. N. Moiseev formulated the forecast of possible consequences of a nuclear conflict between the Soviet Union and the United States (the so-called "nuclear winter" model). To a large degree, this forecast facilitated *START* (Strategic Arms Reduction Treaty), a bilateral agreement between the United States and the Soviet Union on the Reduction and Limitation of Strategic Offensive Arms. The treaty was signed on July 31, 1991 and entered into force on December 5, 1994.

It is possible to differentiate between *active* and *passive forecasts*. A passive forecast is a forecast whose result does not affect (and cannot affect) a forecasted object. For instance, we mention weather forecasts. If the impact of a forecast on a forecasted object might not be neglected (an active forecast [78]), a forecast must then consider the effect of forecasting results. Hence, any normative forecast is active; similarly, descriptive forecasts used in decision-making are active.

2.5 Expertise in Quality Management

The well-known Quality Function Deployment (QFD) method bases on step-by-step multi-aspect expertise. The latter requires accurate work organization for a set of experts, i.e., e-expertise mechanisms. Construction of the customer requirements matrix, transformation of customer requirements into target values for technical descriptors of a final product may include several steps (see Fig. 2.2, where some steps are omitted).

Step 1. Making a list of Customer Requirements for a product. Primary external requirements expressing the needs of customers are specified in the form of second- and third-level requirements. Thus, they form a list of concrete requirements. Wherein not all requirements are known to a customer, expert groups must document requirements dictated by management or regulatory standards. For a market segment, such list may comprise about 50–100 requirements (e.g., maximum speed, body color, comfort level, etc.). Experts have to compile this list of requirements.

Step 2 consists in paired comparison of the importance of different customer requirements by an expert group. This stage ranks customer requirements, i.e., each requirement is assigned some customer importance rating.

Step 3 serves for selecting technical descriptors of new products by experts. Correctly defined target values of these descriptors would meet customer needs stated at Step 1. Technical descriptors are design attributes of a product or service that can be measured against the competition. Later on, technical descriptors must be deployed in specific requirements at different stages of product design, manufacturing, assembly and service in order to appear in the functional performance of new products and customer satisfaction. The list of technical descriptors can be 5 times larger than the list of customer needs (e.g., wearing capacity, robustness, rated power, melting temperature).

In the next step, expert groups have to determinate the direction of improvement for each technical descriptor (this step is omitted in Fig. 2.2).

Step 4 lies in verification of the correspondence between technical descriptors and customer requirements. Here experts analyze the existing relationships between the latter and the former. E-expertise assists in compiling the relationship matrix, where rows stand for customer requirements and columns answer for technical descriptors. This stage may consume much time and involves many expert groups. Actually, experts have to coordinate their actions and generate consentient expert appraisals.

Step 5 promotes innovations. Experts identify inconsistent requirements to new products or equipment (see Fig. 2.3). For instance, "engine power must be improved," whereas "engine weight must be decreased." Such a conflict calls for an additional research work and, accordingly, product redesign and/or production reengineering. At Step 5, it may happen that the list of technical descriptors should be modified or supplemented for adequate reflection of all customer needs.

Step 6. For a new product entering a market, it is necessary to conduct expert appraisal of market characteristics. Such appraisal implies assessing the relative importance of product requirements according to customers (Step 2) and comparing the competitive ability of existing products (customer rating of the competition). Benchmarking takes place. The relative importance ratings of product requirements allow defining the domains of most interest or maximum expectations (on the one hand) and identifying "bottlenecks" to-be-improved. Estimation of the competitive ability of products shows how ultimate users interpret the products against competitors in the sense of their needs satisfaction.

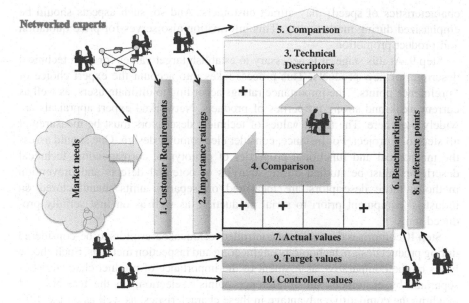

Fig. 2.2 Quality function deployment

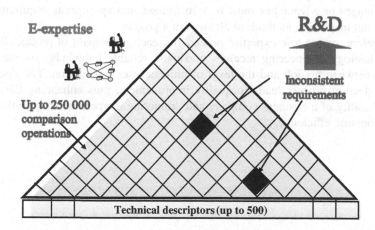

Fig. 2.3 Identification of inconsistent requirements to new products

Step 7 is connected with expert appraisal of the actual values of technical descriptors. Necessary data often follow from measurements and tests. Note that such information covers both the products of a company and its competitors. Experts also rate of the design attributes in terms of organizational difficulty.

Step 8. Using the ranking data in the right-hand columns of the comparison matrix, experts determine the "preference points" of new products. In fact, "a preference point" indicates which aspects in new products (e.g., best-in-class characteristics of speed) may attract customers. And so, such aspects should be emphasized during market entry. This information also serves for price quotation and product promotion.

Step 9. At this stage, it is necessary to establish target values for each technical descriptor of new products. This process takes into account the expert choice of "preference points," the importance ratings according to ultimate users, as well as current weak and strong properties of products. Networked expert appraisals are widely used here. The target values of technical descriptors must be measured at all steps of a project. For instance, consider electromotor design. One should assess the mechanical and functional properties of prototypes; corresponding technical descriptors must be studied during analysis of potential defects and prevention methods; these descriptors are measured on separate units manufactured on industrial equipment priori to serial production, as well as on first serially produced units.

Step 10 comprises the choice of controlled technical descriptors to-be-considered during product design, technology engineering and inspection methods. Such choice bases on the comparative assessment of the importance of product characteristics separated by customers, on "preference points" selection, on the feasibility of reaching the competitive advantage in these characteristics, as well as on the difficulties (opportunities) of target values achievement. Any technical descriptors having a strong impact on customer needs satisfaction and creating competitive disadvantages or advantages must be transformed into appropriate requirements, actions and inspection methods at all steps of a project.

Therefore, owing to e-expertise procedures, each participant of product design and technology engineering receives maximum reliable knowledge on the connection between his job and the level of ultimate user satisfaction. The result of any production process leads to quality improvement, thus enhancing the competitive ability of a business company and promoting the permanent growth of its socioeconomic efficiency.

Chapter 3
E-Expertise: Organization and Technologies

Figure 3.1 demonstrates the institutional organization of e-expertise. In what follows, we discuss it in detail, i.e., provide a list of feasible technologies, study different forms such as polling, electronic brainstorming, etc.

Among relevant factors in e-expertise viability, we should mention the normative legal base, financial provision, as well as historical, cultural and territorial specifics of a certain state. This chapter focuses on some aspects of this kind.

3.1 List of Feasible Technologies

Technologies of e-expertise organization are formed with due regard for the following features. Experts communicate via a network medium, sometimes do not see or even know each other. Nevertheless, they master different techniques of intuitive and logical justification of difficult-to-formalize *skills* such as presentiment, prevision, divining, foretelling, and so on. Exactly these skills and senses appear of crucial importance in e-expertise and decision-making (of course, in addition to unique knowledge and unique experience of experts, a correct comprehension of a current problem situation). We specify several characteristics of e-expertise decisions:

1. the number of experts in decision-making;
2. the significance degree of a decision;
3. the "location" of a problem situation;
4. the degree of urgency in decision-making;
5. the type of assessed object (an area, an enterprise, a report, a draft law, cost price, an event, etc.);
6. the type of expertise object, control level (an authority, a corporate team, a public organization, a family, an individual);
7. the character of cause-and-effect relations and mutual influence of factors;
8. the holistic property of decisions' essence [90], the coverage of all range of factors (with indirect consideration of implicit factors);
9. the degree of problem formalization;

D. Gubanov et al., *E-Expertise: Modern Collective Intelligence*,
Studies in Computational Intelligence 558, DOI: 10.1007/978-3-319-06770-4_3,
© Springer International Publishing Switzerland 2014

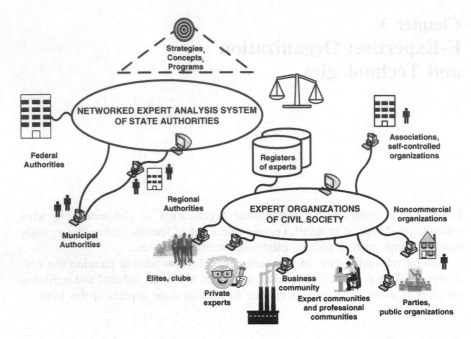

Fig. 3.1 The institutional organization of e-expertise

10. the character of counteraction from an external environment, the dynamics and segmentation of market requirements;
11. the repeatability or uniqueness of a problem situation;
12. the chaotic, quantized, causeless behavior of a problem environment [88], etc.

Depending on these characteristics, networked expert procedures may have different forms of *implementation* [91], e.g.,

- electronic formation of expert comments;
- e-expertise with semantic differential scales;
- e-expertise in the monitoring of current situation;
- electronic brainstorming;
- networked strategic conversation (up to 25 participants);
- networked strategic congress (up to 250 participants);
- crowdsourcing with self-organized networked expert community.

In what follows, we outline certain features of these forms.

3.2 Electronic Formation of Expert Comments

This is a procedure for rapid accumulation of experts' opinions on an issue or a group of interconnected issues, with subsequent analysis of comments, conceptual modeling and substantiation of recommendations for preparing a decision on a

Fig. 3.2 Some aspects of a problem in expert decision-making

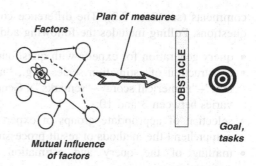

pressing problem. For exhaustive recommendations, experts have to elaborate interconnected proposals on the following structural aspects of a problem: goal-setting, description of relevant factors that predetermine situation development, the mutual influence of factors, obstacles to goals' achievement; measures of goals' achievement and problem solving. Such structure generates the way of a group expert action. Figure 3.2 shows the basic aspects of expert decision preparation. Here the plan of measures can be formally interpreted by some function (a mapping, a morphism) from the domain of factors with mutual influence to the codomain of goals.

Generally, experts express their ideas, conclusions and proposals as a small text, report, review, or reference in a free form. Such comments must be correctly interpreted, comprehended and integrated by analysts. Expert comments can be formed in the following *order*:

- a Principal generates a query to experts on a problem situation;
- an analyst selects an appropriate methodology of expertise conduct and data processing (depending on the specifics of a problem situation);
- the query is refined through several questions according to the selected methodology;
- a moderator administers the expert procedure and collects comments;
- experts give their comments;
- an analyst performs simulation and analytical treatment of comments;
- a generalized expert comment is compiled based on the results of polling.

3.3 E-Expertise with Semantic Differential Scales

Expert polling by an open or closed set of questions using estimation scales (semantic differential scales, linguistic scales) guarantees high-level synchronization and mutual understanding of experts, as well as accelerated automatic processing of the results. Expert polling resembles electronic formation of expert comments (see Sect. 3.2). The difference consists in adding linguistic scales to questions. Polling includes the following additional procedures:

- query generation for experts with questionnaire formalization;
- construction of estimation scales (e.g., "name of estimation scale"–"graduation"–"numerical score"–"remarks"). Generally, the number of graduate marks varies between 3 and 10;
- selection of appropriate groups of experts that master estimation scales and comprehend the methods of result processing (e.g., hierarchy analysis method);
- mailing of the query and estimation scales to experts (with delivery notifications);
- expertise and assessment of the questions by experts, assessment of the consistency index of expert appraisals [99];
- if necessary (in the case of significantly inconsistent appraisals), reformulation of the original query and resending the new query to experts;
- automatic aggregation of the estimation results using certain formal methods (e.g., hierarchy analysis method, averaging method, etc.);
- examination of the expertise results and the aggregated estimate by an analyst;
- automatic selection of the most suitable presentation of the aggregated estimate (according to the analyst);
- meaningful interpretation of the expertise results (possibly, using conceptual modeling).

Expert polling may be accompanied by the following text.

"The procedure of questionnaire filling:

A. Please, put an appropriate numerical score in each cell of the questionnaire according to the estimation scale: (*an estimation scale is given*).
B. The estimate has to reflect your knowledge of the subject (topic) of expertise (instead of your personal attitude to it).
C. For instance, the medium level of some object or subject corresponds to numerical score 3; if you find difficulty in replying, choose "X."
D. For a free-answer question, first suggest possible replies and then assess them by an appropriate estimation scale."

Construction of estimation scales bases on embedded tools of creating and managing *function libraries* and *databases*:

- managing a register of experts with some rating system;
- means of interaction with experts;
- certain techniques for conducting multi-factor analysis in order to assess and forecast situation development;
- a typical set of linguistic (semantic) scales;
- mathematical models for the comparison and cluster analysis of possible alternatives of situation development;
- analysis methods for the causes and consequences of certain factors affecting further development of problem situations;

- some tools of cognitive modeling, hierarchy analysis [99], evolutionary algorithms [91];
- certain elaboration procedures for recommendations on preventing and eliminating the causes of problem situations.

Expertise results demonstration has to provide the following opportunities to an analyst: (1) choosing possible types of estimates' aggregation (averaging method or hierarchy analysis method) and (2) choosing the form of aggregated estimate presentation to a Principal.

Expertise results are presented as experts submit their answers. In case of need, a Principal should be able to study:

- the detailed answers and estimates of each expert;
- the results of analytical processing of expert messages and aggregation of their replies;
- reference information, expert information and analytical information.

3.4 E-Expertise in the Monitoring of Current Situation

We have mentioned that e-expertise in the monitoring of current situation is implemented in two modes:

- solving a specific problem (problem-related monitoring);
- current monitoring of a subject area (permanent monitoring).

Problem-related monitoring gets initiated by a corresponding query of a Principal. Such monitoring results in normative fixation of the fact of problem settlement, completion of the corresponding action plan on a problem or a separate decision. As a rule, in each situation problem-related monitoring calls for a special procedure, simulation method, polling forms and regulations.

Permanent monitoring may be implemented by a Principal or the participants of an expert group on continuing basis. Experts share their appraisals of situation development during given periods. For instance, an expert provides a special-form report at least once a month.

In both modes, an expert generally assesses a situation by a questionnaire. The latter is designed using methods of metric and nonmetric scaling, strategic and conceptual modeling. This assists in automatic aggregation of expert appraisals, prediction of negative consequences of situation development, identification of essential factors and definition of decision-making risks.

During *problem-related monitoring*, one can ask an expert to specify:

- the required control level and control authority for decision-making on an event;
- the degree of urgency in decision-making on an appropriate level of management (an assessment in some estimation scale and period).

Problem-related monitoring can be organized in the following *order*:

- query generation for monitoring;
- construction of an expertise procedure and a conceptual computer model of a problem situation;
- questionnaire design (including linguistic scales);
- selection of appropriate groups of experts (under a constructed expertise procedure);
- mailing of a formalized query to experts;
- expertise of the problem situation by an expert and transmission of an expert message;
- automatic aggregation of the estimation results using the scales and conceptual modeling;
- ensuring Principal's ability to view each expertise result and aggregated estimates;
- automatic selection of the most suitable presentation (visualization) of the aggregated estimate;
- meaningful interpretation of the expertise results using conceptual modeling, definition of decision-making risks, forecasting based on revelation and comparison of different scenarios.

Conceptual modeling requires professional support of specialists in computer simulation, forecasting methodology and intelligent data processing.

During monitoring, a Principal must be able to see:

- the results of conceptual modeling;
- the results of multimedia visualization of situation dynamics on multidimensional scales;
- the results of analytical treatment of expert messages and their aggregation by an analyst;
- the current appraisals of each expert and experts' recommendations on decision-making;
- reference information, expert information and analytical information.

3.5 Electronic Brainstorming

Electronic brainstorming [58] is supervised by experienced *moderators*. This procedure serves for rapid generation of nonstandard ideas and proposals. When experts communicate online and see each other only on computer displays, traditional methods of brainstorming appear inefficient. Receiving text and voice messages, networked experts have to understand each other without ambiguities at maximum shortest time. In addition to correct control of such procedure, it is necessary to involve automatic methods for semantic treatment of numerous messages. They assist in extracting a collective idea from the chaos of text proposals. Here one needs certain methods of nonclassical logics, quantum semantics, and convergent control [88]. A promising approach to electronic brainstorming lies in using different tools of situational awareness and virtual collaboration [58, 84, 115].

Electronic brainstorming can be convergent or divergent. In the former case, experts have to reach a consensus on various ideas they generate synergetically and jointly. In the latter case, experts are stimulated to generate as many ideas as possible and/or most nonstandard ideas.

Electronic brainstorming is the most complicated process of networked expertise in the sense of consensus achievement [58]. For faster consentience among networked experts, it seems reasonable to combine electronic expert procedures with divergent electronic brainstorming based on a technology which guarantees expert process convergence to a consensus. Such philosophy can be represented as the integration of strategic conversations [88], intelligent image processing [40, 116], analytical modeling [91] and virtual collaboration [115]. Different aspects of such integration will be discussed in Sect. 3.6 dedicated to networked strategic conversations.

An ultimate user may participate in the preparation of electronic brainstorming. A moderator has to perform "rigid" control of electronic brainstorming. Actually, electronic brainstorming can be organized in the following *order*:

- query generation for electronic brainstorming and problem statement (e.g., to elaborate an idea, to identify new factors, to assess weak or strong sides of opponents, etc.);
- approximate planning of electronic brainstorming by a moderator including its procedure and regulations;
- rough estimation of possible deviations from the approximate plan of expertise procedure and specification of behavioral logic at deviation points (if possible);
- generation and filling of an associated series of typical questionnaires on computer displays according to the plan of electronic brainstorming;
- selection of appropriate groups of experts;
- automatic choice of abstracts and key words corresponding to the subject domains and publications of experts;
- mailing of notifications on electronic brainstorming (date and time) to experts and online negotiation of their participation;
- electronic brainstorming conduct according to the plan (including questions' formulation and general conduct by the moderator), acquisition of expert messages;
- automatic semantic interpretation of the questions, statistical treatment of the text messages from experts, results' demonstration in a dedicated window during an expert session;
- electronic brainstorming termination by the moderator and preparation of the report.

The plan of electronic brainstorming possibly includes a certain procedure for parallel or successive (circular, time-controlled) acquisition of experts' opinions on a predetermined question. For instance, imagine a two-stage expert procedure (morning session and evening session). In the morning, brief polling can cover the results of "free reflections" by experts, with subsequent construction of a "reflection matrix."

Moreover, the *plan* of electronic brainstorming may stipulate for:

- alternating brainstorming sessions with short duration (e.g., 5 min) and consideration of their results;
- alternating periods of ideas' generation, discussions and collective decision-making on specific aspects of a problem (the duration of each period is 10 min, whereas the total duration of the procedures can reach 3 h);
- successive stages of proposals submission and discussion (without sharp criticism, but with appraisals and complements) and the final stage of decision-making;
- brainstorming with automatic usage of text notifications, boosters and suppressors;
- separate problem-related dialogs with temporal participation of some experts from a group;
- formation of 3–5 criteria for further assessment of different alternatives in expert actions. Such criteria can be sorted by their relevance by the moderator;
- the feasibility of choosing a certain action suggested by experts depending on the state of the procedure, its plan, or other proposals by the moderator. Here it is possible to apply rating systems, semantic scales, interval score, pair comparison and other scaling methods;
- the feasibility of reverting and considering previous stages of the plan, and so on.

In the off-schedule mode, the moderator can "isolate" two or more experts for faster mutual understanding or reaching a local consensus on some issue. We have emphasized that the most efficient technique of consensus achievement is the combined application of analytical tools of virtual collaboration [115], including implementation of the following components: text and voice recognition for experts, construction of text histograms and clusters, evaluation of expert participation indices, and real-time visualization of the results.

The process of electronic brainstorming must be displayed and available to all participants (except individual messages). Experts provide voice messages, various drawings and figures, smileys, etc. However, the amount of any supplementary materials should not exceed 1–2 pages per expert. This requirement guarantees that all participants have time to study such materials within the limited bounds of electronic brainstorming. In the case of voice messages, it is necessary to perform their automatic analysis with separation of most stressed words.

Electronic brainstorming can be organized, e.g., by the *blackboard* methodology. Everybody can imagine the "interior" of a classroom with a blackboard, chalk and damp cloth. In the case of networked implementation, we obtain the following framework. The blackboard becomes electronic, any records can be stored on a computer and displayed when necessary, any corrections admit logging in history files (Edit mode). The blackboard model serves for conducting sessions from a situation room–all records can be displayed for collective usage, and all participants can access networked dialogues.

Electronic brainstorming requires general problem statement by a Principal and a moderator. Such statement possibly comprises several components, namely,

- the goals and tasks of brainstorming (convergent or divergent brainstorming);
- a brief description of a problem situation with possible sources of additional information;
- an intuitive forecast of situation development, any misgivings of a Principal;
- the desired trends of situation development, a certain approach to situation development planning, etc.

During further preparation and conduct of electronic brainstorming, these components must be characterized not just on the descriptive level, but on a formal mathematical language (e.g., a language accepted in applied systems analysis and modeling [7, 91]). Problem statement should employ the following *groups of participants*:

- managers with the authority to sign, subject supervisors, experts and specialists in a particular problem, as well as experts in subject domains (if dictated by the procedure);
- moderators, cognitivists and knowledge engineers (specialists in formal description of different problems, who construct a "bridge" between content and forms, Principal's comprehension of a problem and the corresponding computer model);
- mathematicians and program developers.

3.6 Networked Strategic Conversation

A networked strategic conversation can be conducted under Principal's chairmanship. Just 2–3 networked moderators drive such conversation, if the number of participants does not exceed 20–30 persons. A networked conversation possesses strategic orientation, i.e., it implies consentient goal-setting, problem formulation and elaboration of future actions. Owing to special conditions, all distant participants are able to reach mutual and unambiguous understanding as fast as possible by exchanging of their messages and multimedia visualization tools. For better understanding, a moderator involves certain tools of virtual collaboration [58, 115]. Generally, networked strategic conversations engage team members leaded by a Principal.

Networked strategic conversations intended for settlement of different problems posed by a Principal have organization procedures basing on the following *methods*:

- strategic situation analysis methods with proper consideration of socioeconomic priorities and assessment of most important factors resulting from independent brainstorming [31, 53, 61, 88];
- hierarchy analysis and cognitive modeling methods that ensure strategic forecasting by quantitative and qualitative factors [2, 94, 99];
- methods from quantum semantics and evolutionary computations that account for existing "implicit" opinions of experts, the feasibility of considering paradoxical interpretations in modeling, as well as modeling based on future vision in the absence of past data [4, 46, 91, 111];
- solution methods of inverse problems, methods from catastrophe theory, thermodynamics and convergent control [88, 90];
- latent solution synthesis methods (in addition to the context of expert messages, they use accumulated data on electronic traffic during expert communications);
- traditional social study methods including organization of focus-groups and depth interviews;
- traditional statistical modeling and extrapolation forecasting methods;
- evaluation methods of strategic correction coefficients for statistical forecast amendments (based on modeling), situation development prediction with proper consideration of strategic factors, and so on.

The typical moderation order of a strategic conversation comprises the following *stages*:

1. Constructing a weighted tree of goals in problem solving (e.g., by hierarchy analysis method [99]).
2. Identifying external and internal factors which characterize situation (e.g., using SWOT analysis).
3. Making a priority list of problems impeding goal achievement (based on matrix correlation analysis and generalization of external and internal factors).
4. Making a list of action perspectives (directions) on the basis of revealed problems.
5. Assessing the priority directions of actions and the impact of different factors on situation development (e.g., via cognitive modeling and genetic algorithms [90, 91]).
6. Preparing well-grounded recommendations and draft decisions.

This moderation order can be described in greater detail.

A. All participants are sent information on the goals and tasks of a strategic conversation in a certain format with requirements specification. All participants are notified and the strategic conversation is organized.

B. For establishing a meaningful contact with participants, the moderator asks 1–2 questions, e.g., "Please, tell me in one sentence, how you assess the situation," "What is your desired situation after a year?"

C. Preliminary goal-setting is conducted for problem solving. Goals can be represented by a *tree of goals*. For instance, experts compile three questionnaire tables in the networked mode: (1) "Primary goal (mission)," (2) "Second-level goals whose achievement is desired, but they lie beyond Principal's and team members competence (up to 10 goals)," and "Third-level goals whose achievement is in the competence of the Principal and team members (up to 20 goals)."

D. Questionnaires filling with proper identification of internal and external *factors* of the problem. External factors have slight dependence on a Principal and team members, whereas internal factors lie in their competence. All factors of the problem situation studied by experts must provide its complete coverage [90]. The total number of factors may reach 100. Furthermore, factors can be classified as positive (favorable) and negative (unfavorable) ones. Factors may represent:

- quantitative indicators (temporal series, statistical data, numeric expert appraisals);
- nonquatitative indicators (mainly qualitative and conceptual definitions, expert information).

E. Expert polling serves for assessing the interconnection and interaction of factors. The *interconnections of factors* are the impacts of one factors on others (generally, estimated by some numerical score). Here one can employ various visualization aids (interconnection graphs or matrices). An arc connects two nodes of a graph as follows. The beginning of an arc corresponds to the factor having an impact on the factor at the end of the arc. This technique enables analyzing each pair of nodes (factors). Two multidirectional arcs may exist between two nodes, either. A factor possessing no connections with other factors should be reanalyzed.

F. Given a set of factors, joint efforts of all experts (their polling) separate certain factors:

- which lie in Principal's and team members competence (internal factors, also called *managerial factors*, e.g., appointment of subordinates, wage rise);
- whose values should be modified rapidly, but this task seems impossible for a Principal and team members (*external factors*, e.g., market requirements).

Each action mentioned may result from certain measures or projects on reputation or organization improvement, advances in economy or normative and legal base, etc.

G. By expert polling, feasible managerial actions on different factors are selected. Possible *scenarios* of actions are outlined.

H. The above scenarios serve for computer simulation of situation development. This stage assesses the impact of managerial factors on goal factors (in the course of time).

I. While puzzling over different questions like "What should be done to...?", one can apply evolutionary computations (the genetic algorithm) [91]. Actually, it rapidly defines the optimal force distribution among team members and interaction of managerial factors, thus ensuring the synergy of actions for goal achievement.

J. For getting the priority list of problems impeding goal achievement, experts fill the *window* (*matrix*) *of opportunities* (SWOT analysis). The rows and columns of this matrix match the internal and external factors of the problem. Each cell contains the expert appraisal of the relevance of combining appropriate factors for problem solving. For instance, imagine that some threat factor turns out relevant for a Principal's strength. In this case, the expert appraisal of such factor combination becomes high. It is possible to use decimal scales from −1 to 1.

For the current state of an external environment, such analysis allows defining the relevance of (a) the strengths and weaknesses of activity and (b) the threats and opportunities for management efficiency improvement.

The relevance levels of different factors are assessed (by summing up expert appraisals in each row and column) taking into account the mutual influence of all factors. These levels can be used for choosing some factors for further processing.

K. A priority list of action directions (*priorities*) is made. This would lead to consentient reformulation of the priority list of problems.

L. According to the priority action directions, experts form the lists of *measures* and *projects*. Generally, they correspond to the factors used to choose the priorities.

M. Recommendations based on the results of collective expertise are stated by the subject analyst of a direction (subject supervisor) in the form of a note, a concept, etc.

All information synthesized by a networked strategic conversation serves for elaborating a concept, a strategy, a doctrine, and so on.

3.7 Networked Strategic Congress

This e-expertise procedure is a natural extension of networked strategic conversations. The number of direct participants of a *networked strategic congress* may reach 250. Organization of such congresses possibly involves over 25 networked moderators.

A strategic congress on a specific subject can be conducted in three modes, viz.,

- all participants are located in a same building or room;
- geographically distributed experts have networked interaction;

- the mixed mode when some participants share a same building or room and the others are available through communication channels.

In fact, only the second and third modes represent a networked strategic congress. Its implementation presumes solution of the following tasks:

- regulated preparation of strategic congress conduct (including multimedia, scenarios, commentaries and overall direction). The period of preparations may constitute approximately 3 months;
- formation of electronic activity support cyclograms for different-form groups of participants: conferences (a regulated forum of all participants), strategic conversations (see Sect. 3.6), and problem-related groups (about 10 participants);
- provision of the communication activity of participants and their groups using e-expertise tools, with acquisition of communication data and traffic assessment of experts;
- informational support for the motivational control of specific participants and their activity;
- parametric adjustment and adaptation of the technological environment intended for facilitating interaction in different types of collective expert procedures;
- strategic congress coordination including regulation of different events and interaction among subject supervisors, moderators and expert groups;
- further treatment and generalization of the results using conceptual modeling tools and organization of collective expert procedures.

A networked strategic congress represents a rather new electronic mechanism of strategic control in socioeconomic systems with well-timed conciliation of interests pursued by different public subjects (citizens, societies, corporations, regions, authorities, etc.). Here one can employ special techniques for accelerated reaching a consensus of all congress participants on desired goals and necessary means.

3.8 Self-Organization in Networked Expert Community

The matter concerns the mechanisms of expert's *self-organization*. Interestingly, the process of such self-organization is affected by Principal's intentions ("from above") and the expectations and needs of population or a dynamically segmentable market ("from below").

Self-organization favors fast revelation of original ideas, factors, goals, and proposals, as well as appreciable reduction of risks and prevention of the negative consequences of managerial decisions. Exactly self-organized environments bear networked leaders and identify talented moderators and experts (see Chap. 4 of the book).

This task is mainly solved within the informational field of Internet. It includes the following *subtasks*:

- automatic revelation of websites and portals which conduct forums, acquire expert information or organize discussions of similar problems;
- content analysis of text messages of experts during networked discussion of issues (with revelation of positive and negative features in expert judgments);
- building the archetype structure of subjects and problem situations, content dynamics tracking for websites and forums by intelligent robots;
- automatic revelation of ideas, factors, goals, and proposals facilitating solution of specific tasks, risks reduction and elimination of negative consequences of managerial decisions;
- connect analysis of interaction among experts (message traffic analysis);
- automatic revelation of talented moderators, experts, leaders, etc.

Intelligent robots have to track the group dynamics of discussions on a studied subject. Here *intelligent information technologies* and certain methods of *content analysis* of data flows assist in extracting useful and meaningful information.

To find expert leaders, talented networked moderators and organizers, it is possible to track the life cycle of an expert group using e-expertise technologies (see Sects. 3.2–3.7). This life cycle includes several *stages*:

- at the *initiating* stage, a problem situation gets formed (e.g., anxiety in some section of a society, which creates prerequisites for positive and negative opinions of people). There arises the latent need for opinions' integration and expression of the unanimity of this section;
- at the *pregroup* stage, a group of experts for opinions' coordination is still not outlined, but there exist separate opinions of experts on a specific subject. This stage can be crucial for further efficient operation of a self-organized expert group. This stage serves for reaching a consensus regarding the individual goals of group members, as well as for revealing the motivation for group expert activity. Exactly this stage manifests the behavioral attribute of group development, which characterizes the orientation towards competition in a group (divergence) or towards mutual assistance (convergence);
- the *early* (initial) stage of expert group formation covers the period of comprehension and analysis of a problem situation. On the one hand, group members endeavor to find their place in this group intuitively and logically (by thinking of their benefits from future collaboration with other group members). On the other hand, they define the goal and functional orientation of group activity. One can observe the appearance of a leader predetermining the convergence orientation of group development;
- during the *transition* stage of expert group formation, the participants realize and study possible risks, anxiety, the need for safety, the feeling of resistance, conflicts, control and other possible problem situations. The prominent role at this stage belongs to interactive information technologies and methods of reaching a consensus for assessing the degree of resistance to expertise. The

above technologies and methods must guarantee an adequate response to possible delays in replies, the content-richness and tone of different messages and, possibly, to visual and symbolic characteristics of behavior, etc. In fact, such technologies and methods are introduced by the leader; this explicitly raises the trust of other group members in the leader;

- at the *working* stage of expert group formation, one can see a consolidated consensus of the participants on the goals and ways (priorities, projects, measures) of their joint actions, as well as implementation and control of such actions. Again, the technologies and methods suggested by the leader facilitate these processes. This makes the activity sensible and motivates each participant to bear personal responsibility for the result. The working stage implements the control process of group activity, reveals leadership and the talent of the leader acting as a moderator;

- the *concluding* stage (which comes sooner or later for an expert group) terminates group work. Withdrawal of each participant from a group represents an event to-be-assessed analytically and semantically.

The content analysis of expert information. Many issues under expertise require thorough and well-substantiated answers. And so, one would hardly depreciate the problem of text processing for expert opinions. The topicality of automatic text processing methods grows with the number of objects under expertise and the expansion of expert communities. As the volume of text information from experts exceed some threshold, its manual treatment becomes an impossible-doing task. This makes the idea of distributed (networked) expert communities infeasible.

The problem of automatic analysis of text information is solved by modern linguistic and statistical methods implemented in the content analysis module.

This module extracts informational objects from texts and saves recognized objects in a data warehouse for further statistical treatment. We mention the basic functions of the content analysis module:

- tone analysis (positive/negative) of experts' opinions on some objects or subjects, events, decisions, etc.;
- problem profile identification;
- separation and analysis of problem causes;
- revelation of implicit connections and constellations of experts;
- classification, clusterization, categorization;
- statistical analysis and visualization of the results of linguistic processing of texts, etc.

The content analysis module can be efficiently used to acquire and analyze expert information from various sources (e.g., Internet). Concerning the completeness of the resulting appraisals of an issue under expertise, we emphasize the role of automatic analysis of the so-called blogosphere (personal websites and weblogs on certain issues, considered collectively).

The results of the meaningful exploration of texts performed by the content analysis module can be employed:

- to define the opportunistic and conformist behavior of experts;
- to organize cognitive mapping and modeling for experts' support;
- to construct typical cognitive models of subject domains, etc.

The connect analysis of experts' interaction (*traffic analysis*). Communication of people by words covers bounded volumes of information. During a personal contact, much information is expressed through intonation, mimics or gestures. Sometimes, words and signs mean nothing for the participants of a meeting. To a large degree, the success of any organization depends on the atmosphere of trust and emotional attraction of employees (rather than on printed or pronounced words, signs, instructions or strict regulations).

Even the fact of messaging (without content analysis) can be important. The fact of message transmission generates a commitment. Certain interest consists in analyzing the *facts of messaging* among different authorities or companies obeying certain regulations. The facts of message transmission initiate communications, thus producing explicit and latent information.

Latent information concerns everything which seems ambiguous, unreliable or implicit. For instance, we mention the sense of trust, the state of stress, the creative level of an interlocutor; the reason of irregular supplies; the hidden meaning of a text message; the implicit relationship of events; the emotional potential of a company; the motivation and interests of employees; product quality, the characteristics of sole rights, the vigilance of team members, the success appraisal of a risky action.

The presence of latent information causes problems in designing control mechanisms for decision-making on company management. Especially, when such control requires information processing without quantitative estimation. The complexity of these problems grows with the volume of information, the sizes of a telecommunication environment, the number of participants of electronic communication.

Rapid, visual and convincing replies to different questions such as "What should be done to reduce financial risks?", "How will a certain measure affect the business reputation of an organization?" construct an experimental model of a situation and provide appropriate information.

In the context of expert procedures, there exist several restrictions of conventional models employing words, formulas and schemes. First, most problem situations could not be described by quantitative or even qualitative indicators. Second, mathematical transformations may fail to guarantee the true ultimate result. Third, managers of organizations and employees often trust the models developed only by themselves. And fourth, message content analysis sometimes can be impossible due to technical, temporal, corporate or normative and legal restrictions.

In such cases, one takes advantage of different traffic analysis tools for electronic messages without their content analysis and assessment. The scheme of a corresponding analytical process is illustrated by Fig. 3.3. Here readers observe an example of a certain technology of situation monitoring. The flows of event

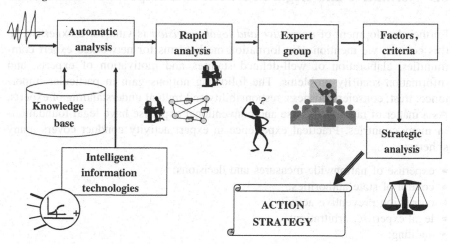

Fig. 3.3 The scheme of latent monitoring

descriptions and electronic message traffic are the inputs, whereas an action strategy makes the output.

Automatic processing of the input flows distinguishes "*problem situations*" for further expertise. Expert analysis identifies factors characterizing a problem situation, and strategic analysis yields an action plan.

Monitoring of electronic messages allows:

- to draw the integral picture of connections among all participants of electronic messaging (experts);
- to reveal the interconnection between the activity of message subjects and specific economic, political or social situations;
- to forecast the activity of legal entities in execution of governmental orders;
- to recognize stable structures with system organization and assess the role of specific subjects in their activity.

Traffic analysis software for electronic messaging of experts may include:

- a database maintenance system for the events of electronic messaging among the participants of a group expert procedure;
- a standard content analysis system for electronic messages and related information (e.g., from Internet);
- a selective connect analysis system for the traffic of electronic messages using intelligent information technologies (neural networks, genetic algorithms, cognitive analysis, and so on).

Yet, message traffic analysis has to follow existing legal regulations including the area of personal data management.

3.9 Normative and Legal Provision of E-Expertise

Further development of *normative and legal provision* is vital for e-expertise. In this context, we mention new formation mechanisms for networked expert communities, elaboration of well-defined statuses and motivation of experts, and information security problems. The following notions gain in particular importance: trust, conscientiousness, responsibility and mutual understanding of experts. As a matter of fact, e-expertise and conventional expertise have legal foundations in many countries. Practical experience in expert activity conduct covers many spheres:

- expertise of nationwide measures and decisions;
- control of state authorities;
- corruption-preventive activity;
- legal expertise, arbitration;
- auditing;
- legislature;
- the activity of self-regulated organizations;
- industrial expertise, etc.

To all appearance, further development of the normative and legal provision of e-expertise calls for establishing a more precise correlation of several notions (see Fig. 3.4). The latter include conscientiousness. It can be acknowledged as a principle of civil law. The concept of conscientiousness must be applied to assess the rights and duties of different subjects in expert activity.

In expert activity, one can comprehend conscientiousness as a morally acceptable behavior of different participants of expert procedures, which agrees with existing public beliefs of integrity, utility and harm. How can we assess the ethicality of expert's actions in a specific situation? Formal attributes for such assessment may include certain information on expert's awareness regarding circumstances taken into account during his decision-making. Other attributes can be used to judge the intentionality or involuntariness of expert's actions. Here it seems rational to elaborate *Professional Ethics Code for Networked Experts* (see the Appendix).

To be efficient, an expert has to bear personal *responsibility*. Maximum centralization must take place in the methodological and practical guidance of expert activity. Quite the contrary, the responsibility for contribution to problem solving requires maximum decentralization. To a large degree, the responsibility of experts is predetermined by their legal status, involvement in real decision-making, commitments and, accordingly, by the share of responsibility (material, status or moral responsibility) experts bear *de jure*. Experts can make mistakes, e.g., underrate an inconspicuous factor or overrate a common factor.

Sometimes, an expert makes recommendations (ideas, proposals, or considerations) and bears no responsibility for their implementation. According to a contract, an expert may have certain *obligations* to compensate possible losses due to

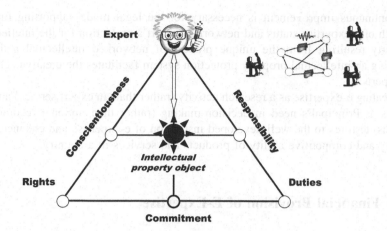

Fig. 3.4 The correlation of basic notions

planned risks or bear financial liability and property liability. However, this surely reduces his creative activity.

Anyway, an expert possesses *social responsibility*. There exist several types of social responsibility in expert activity (legal, professional, reputation-related social responsibilities, to name a few). All these types require clear regulations in expert activity. For instance, law-breaking is a ground for legal responsibility; in this case, an expert is punished or penalized. However, any threat of punishment produces fear and redoubles caution in generation of expert ideas. Such negative approach to responsibility may exist owing to the development of legal mechanisms of lobbying and anti-corruption drive. It seems that the idea of positive responsibility corresponds to the principle of conscientiousness in expert activity.

The completeness of copyright or *intellectual property* rights for all materials provided by experts generally belongs to experts (except cases with special mention). When information is provided by experts with certain compensation, intellectual property rights for expert products generated as the result of expert procedures can be transferred to a customer (subject to mutual agreement). This situation is possible in cases when an expert registers his intellectual property according to existing rules (patents, publications, etc.).

Information technology allows documenting the ideas and proposals of experts, which enhances the capabilities of intellectual property protection. Meanwhile, operation in information space presupposes legal support for many aspects of expert's intellectual property protection, viz.,

- the results of intellectual activity at the expense of state budget;
- the fees for enjoyment of company's services, assets and intellectual rights;
- disposal of a joint intellectual property object by a rightholder, etc.

Continuous improvement is necessary for the legal mode supporting further growth of e-expertise status and networked expert's protection for the intellectual property resulting from the unique process of networked intellectual activity. Refining an intellectual property protection system facilitates the creative activity of experts.

Treating e-expertise as a research activity (rather than an expert service) and its results as Principal's need in decision-making (rather than optional recommendations) testifies to the well-developed institution of e-expertise and enhances the quality and competitive ability of products and services in a country.

3.10 Financial Provision of E-Expertise

Generally, the existing *financial mechanisms* and legislature of a country do not ensure efficient operation of an expert community. It is necessary to adopt special approaches. In addition to experts cooperating with state authorities by contracts, there are other groups of experts:

- striving for their personal or collective interests;
- lobbying the interests of science-consuming companies and various business corporations;
- representing the interests of intellectual clubs, public organizations, networked communities and other civil law institutions.

International experience shows that, within the existing typology of contracts, increasing the efficiency of expert activity calls for separating a special type of contracts, i.e., contracts for expert services and expert research work (hereinafter referred to expert services). The underlying reason is that *expert services* appreciably differ from other services. We elucidate these distinctions below. First, an order for an expert service may appear spontaneously and rapidly. Second, substantially greater uncertainties arise in assessing many parameters such as intellectual property, effectiveness, reliability, creativity, utility, riskiness, responsibility, conscientiousness, independence, trust, emotionality, rating assignment, etc.

Financial mechanism design for expert activity should be focused on:

- separating a dedicated branch in law (federal expert contract law);
- guaranteeing the responsibility of state authorities for expert information usage;
- creating and ensuring fiscal stimulation of independent sponsor funds for expert activity;
- share participation of a state customer and an expert company (a contractor) in reimbursements of costs and risks;
- precise regulations of expert selection (including legal entities) for federal contracts;

- allocation of intellectual property;
- increasing the prestigiousness of a state order for expert services (the large scale, predictability and profitability of state orders);
- the feasibility of using a federal expert contract as a guarantee (pledge) tool;
- making expert's e-cards for identification and payments, and so on.

Moreover, the formation mechanism of state needs in expert information must be defined on the basis of several information flows–from state authorities, scientific organizations, business communities and other participants of expert activity. Thereby, all participants of expert community get involved in the process of making state and corporate decisions. This naturally increases their motivation to expert activity.

3.11 Motivation of Experts

The development of networked expert activity permanently improves the *motivation of experts* (especially, nonfinancial motivation) through the following components.

Participation in managerial decision-making. The extended rights and raised responsibility of experts enlarge their role in elaboration of normative documents, improve the demand for creative work, as well as reduce unemployment among brain workers. This legitimizes the lobby for small- and medium-size science intensive business companies, heightening the competitive ability of a country, the quality of products and services, the quality of life and tax levies in a state budget.

Growth of personal reputation and abilities. Participation in expert community operation promotes the objectification of expert activity assessment, improves the illustrativeness of their results and involvement of experts in a real economy. The advantages and shortcomings of an expert reflected in the appraisals of his work increase the transparency of expert institutions and perfect the public image of experts.

An expert has to accumulate related knowledge and deepen the comprehension of problems in a corresponding subject domain. This intensifies competition, and an expert faces difficulties to attend different expert events conducted simultaneously. In such conditions, networked and intelligent information technologies assist in collective synthesis of new knowledge (this task can be impossible for an individual expert).

Rating assignment to experts. Motivational role also belongs to well-timed qualitative assessment of labor efficiency demonstrated by expert community members. This is possible in a certain rating system of experts. For rating assignment, different characteristics of expert's activity can be classified in three blocks:

- the holistic approach to problems;
- nonfinancial active, emotional potential;
- management of expert procedures and knowledge.

Acquisition of exclusive analytical information. An expert obtains access to exclusive and reliable information from federal and local authorities, including information from corporate portals of different ministries and departments, forecasts and results of group expert procedures.

Free analytical and computational services. Expert community members can enjoy various services on informational and technological support of their analytical and predictive activity. Commercial application packages for analytics may cost tens or even hundreds thousands USD. There exist exclusive methods and technologies not available on a market. Here we mention organization methods of strategic conversations and congresses, electronic brainstorming, polling by semantic scales, intelligent information technology, as well as many acceleration techniques for creative activity (solution algorithms for invention problems and complex inverse problems, cognitive modeling methods, logical and heuristic methods, etc.). Expert community members can use all these tools in groups or via networks (*cloud computations*) and receive professional support if necessary. They do not purchase expensive software licenses.

Accreditation, certification and decorations Accreditation of an expert can be performed by authorities, corporations, companies or other organizations according to an accepted procedure. An expert is included in the register of experts, which raises his reputation potential. Expert community members may receive the certificates of international-level experts or state authority experts, badges and decorations, special membership cards, etc.

Protection of expert's rights and freedoms is implemented by further institutionalization of expert activity, viz., creation of noncommercial and self-regulated organizations, labor union formation, and so on.

3.12 Training of Experts

Any expert possesses certain knowledge and skills for networked activity. Nevertheless, networked operation of expert groups requires additional knowledge on synchronization of collaboration, methodological support and improvement of mutual understanding. This is the task of *training of experts.*

The amount of training depends on the role of a specific expert in expert activity (a government official, a corporate manager, a supervisor, a group leader, etc.). The minimum period of training varies between 8 and 72 h. Certain knowledge can be updated annually (e.g., in the field of legislature).

Additional training admits distant forms. The most important topics of training concern organization of group networked analysis of different situations, consentient elaboration of managerial decisions, and conduct of networked expert procedures. The program of such training may include material from different disciplines such as philosophy, psychology, political science, sociology, mathematics, physics, management science, solution of inverse problems, controlled chaos, quantum semantics, integrated mechanisms of organizational behavior control and others [24–26, 57, 88, 91].

All study guides and textbooks can be available on expert community portals. An expert may receive periodic notifications to read some novel information. On rare occasions, an expert is asked to solve certain methodological problems or settle didactic issues through questionnaires.

Chapter 4
Trust Networks and Competence Networks

Expert finding for specific e-expertise is a multidisciplinary problem at the junction of strategic analysis, decision-making theory, synergy, inverse problem solution, human capital and emotional potential assessment, motivation control, quantum semantics, knowledge management, organizational analysis, information retrieval (acquisition, indexing, storage of artefacts and other evidence of expert knowledge), and analysis of social and organizational networks. It seems often difficult or even impossible to localize potential experts in an organization. The reasons consist in very many employees of an organization, its functionally and/or geographically dispersed character. What is more important, problem solving may require inviting experts from related or totally different subject domains.

The need for expert finding for e-expertise can be caused by the following. First, it is necessary to answer a given question (without discussion or long-term communication). Second, it is necessary to find specialists with desired skills (with discussion and explanations). Third, it is necessary to accumulate information on a circle of potential experts in certain subject (expert selection for future collaboration). Fourth, experts wish to share their knowledge. And so on.

Here an essential aspect concerns the simultaneous superposition of several networks:

- *trust networks* (nodes are agents, arcs are relations among them, i.e., trust);
- *competence networks* (*nodes* are the set of competencies and the set of *agents* (potential experts), and *relations* (*arcs*) are the correspondences between agents and competences);
- *organizational networks* (nodes are agents, relations are institutional structures, organizations or organizational roles and/or positions, subordinance relation, etc.);
- *topical* (*subject*) *networks* (nodes are agents, relations are classes of discussed problems);
- *communication networks* (nodes are agents, relations are different communications, e.g., direct or indirect communication, etc.) and others.

D. Gubanov et al., *E-Expertise: Modern Collective Intelligence*,
Studies in Computational Intelligence 558, DOI: 10.1007/978-3-319-06770-4_4,
© Springer International Publishing Switzerland 2014

The unifying model is the concept of a *multinetwork*, where nodes correspond to agents, competencies, subject or problem domains, roles, positions, etc., and multiple arcs among them characterize some relations, i.e., connections between experts and their roles, competencies, and so on.

This chapter surveys modern approaches to definition and revelation of the *set of potential experts*. In addition, it explores certain methods of restricting this set to the *set of experts* participating in specific expertise. Of course, such methods depend on the goals pursued by a Principal.

4.1 Experts in Business Processes of Organizations

To concretize finding methods of a required expert, we should describe the *context* of his activity. Generally, an expert represents a competent specialist in some domain; being an employee of a certain *organization*, he participates in its *business processes*. During a business process, a specialist (an expert) interacts with other employees of the organization. Making decisions within his competence, a specialist implements the following cycle: observations of a current situation; assessment of a current situation; planning and decision-making; execution of decisions [36]. At each stage of this cycle, an expert has to apply his experience and knowledge from different *sources* (if any) or create new knowledge.

Possible sources of knowledge are *people* (individuals, groups, teams) and various *artefacts* (documents, informational systems).

Ready-made knowledge for a problem being absent, one should generate new knowledge. Imagine that a problem has been formulated. On the one hand, an expert studies the problem independently (by accumulating the existing professional knowledge on the problem and analyzing the existing materials in the context of this problem). On the other hand, all ideas are intensively discussed by a competent group during study (the ideas of individuals are verified through group discussions and debates on traditional and/or networked seminars, conferences and conversations).

Knowledge storage in the *database of an organization* (for further usage) requires knowledge codification in the form of conceptual models, cognitive schemes, ontologies, frames, artefacts, etc.

As the final result, professional knowledge enters different databases (journals, annual reviews, scientific publications, etc.) in the form of "papers." In other words, they are translated in a community via different resources and/or used to learn new employees.

Searching for evidence of expert knowledge. The aforesaid implies that information on the professional activity of an expert can be found as follows, see Fig. 4.1 An *expert* (e.g., an expert researcher):

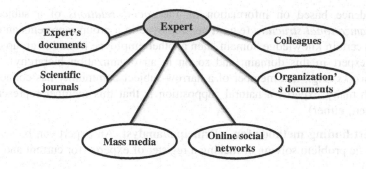

Fig. 4.1 Sources of information on experts

(a) *participates* in project work (which produces different documents connected with a project and an expert-reports, requirements specification and other project documentation);

(b) *creates* working notes added in inverse chronological order during professional activity (live journals, blogs);

(c) *participates* in numerous working events (conversations, seminars, etc.) which produce different documents and artefacts (records, video and audio materials);

(d) *trains* new employees (seminars and courses on professional development) and answers questions (correspondence, e-mailing, forums, Q&A services);

(e) *uses* information/knowledge in his activity (appeals to colleagues and other sources of information);

(f) *creates a profile* with information on his activity (CV, employee directory, "yellow pages" in Web, Wikipedia, professional social networks);

(g) *participates* in different groups and professional communities (e.g., a research group of a supported project and other independent groups (Google Groups, sourceforge.net), an organization (a department, a division, a working group), communities of interest, professional (expert) communities, associations and councils);

(h) *publishes* papers in different authoritative resources and *is cited* by colleagues;

(i) *appears* in news (discoveries, awards) and *announces* his professional opinions on different sites.

Evidence of expert's knowledge. We have mentioned that artefacts or social subjects represent possible sources of knowledge. And so, evidence of expert's knowledge can be divided into two categories:

(A) *Artefact evidence* for expert finding: *structured* (database entries, CV, employee directories), *semi-structured* and *unstructured* (e-mail messages, work programs and action plans, memoranda, records and log journals, blogs, images, audio and video materials). Artefacts can have different types, either (systems, tools, documents). We consider artefacts as data objects indicating of expert's competencies.

(B) Evidence based on information on the *social relations* of a subject: an *organizational structure* (e.g., if an employee in a group represents an expert in a certain knowledge domain, then another employee from this group can be an expert in this domain; and so on in an organization hierarchy); *social networks* (e.g., if a member of a narrow-subject community has connections with many experts, a natural supposition is that this member represents an expert, either).

Expert finding methods for a Principal/analyst. An expert can be searched for specific problem solving or making a register of experts (for current and future projects).

Generally, experts' selection for specific problem solving takes place from an existing register of experts. Such registers include certified experts (the ones passing special accreditation). As a rule, this is caused by time limits and, above all, by the need for choosing experts from an established trust space. Yet, a specific problem may require new experts.

We emphasize the following ways of new experts' retrieval:

(active search)

- official inquiries (e.g., addressed to the CEO of an organization);
- inquiries at online forums, Q&A services, mailing lists;
- recommendation inquiries to acquaintances (possibly, by chain), employers, experts (using mutual recommendations, sequential recommendations, etc.);
- invitations (e.g., Google ads), tournaments and tests;
- an expert seeks for "customers" himself.

(passive search)

- application of search engines (Google, Yandex);
- analysis of registers (e.g., employee directories of organizations), analysis of belonging to professional communities, communities of interest, communities of practicians;
- study of project documentation, guides, requirements specification and other corporate documentation;
- study of publications from authoritative sources (analysis of opinions of authoritative sources);
- study of correspondence (e.g., among employees within discussion posts and mail);
- search by indirect attributes (e.g., taking into account the vertical and horizontal ties of users).

4.2 Expert Finding Problem

Many problems connected with expert finding presuppose answering the following *questions*:

- Who possesses required competencies for solving a given problem? Who possesses related and additional competencies for solving a given problem?
- Which competencies are inherent to a given expert? Which expert is similar to a given expert?
- Which competencies are inherent to a given expert community? Which expert is most significant in a given expert community?
- What is the emotional or creative potential of an expert?

We consider expert finding as a complex problem comprising several *stages* [3]:

(1) *query formulation* (definition of necessary competencies for problem solving and choice of an expert search strategy);
(2) *revelation of experts* (formation of a certain register of experts and their ranking);
(3) *selection of experts* which best match specified criteria. This stage takes into account expert's rating, availability, location, the role in an organization, and current tasks fulfilled by the expert. Some assessment tools for expert ratings can be found in the Appendix.

It seems that, in addition to the competence aspect, expert search should also give proper consideration to the *social and temporal aspects* of expert's activity. The social aspect implies studying various factors such as personal qualities of an expert (communicability, independence, the ability to render assistance); trust in an expert and his reputation; relations among experts (strong or weak ties, horizontal, vertical or diagonal ties, personal relations and working relations, professional contacts and common interest relations); information flows among experts (social relations are important for maintaining awareness, knowledge and competencies). The temporal aspect implies studying other factors such as the evolution of competencies; education, current place of employment, current projects and completed projects, the validity of expert's knowledge, the awareness on current competencies.

Note that the expert finding problem can be treated in a wider statement (instead of the best expert, search for several experts to form a team). Here one takes into account the subject (topical) profiles of experts (their competencies) and the social profiles of experts (their social relations and compatibility with other users) [14]. Formally, a good team of experts must possess all competencies needed for solving a given problem (at least, one expert per each competence). On the other hand, it is desired that all team members form a social network subgraph with good connectivity [59].

4.3 Expert Finding Models

Originally, expert finding systems used *competence databases*. They were updated manually, by system administrators or users (e.g., during registration) possessing expert knowledge. Therefore, experts could be retrieved through database queries or browsing the directory of experts ("Expert register").

Notwithstanding its rapidity and ease, such approach suffers from a series of shortcomings. First, people are inclined to self-partiality or negative narcissism; the corresponding foundations relate to cultural and/or national traditions, striving to meet the expectations of other people, the need for information conceal due to confidentiality, etc. Second, this approach consumes more efforts for database initial setting up, loading and maintenance. Third, several methods of e-expertise require adding new individuals to experts (e.g., networked strategic conversation, electronic brainstorming). Therefore, modern conditions facilitate the appearance of expert finding systems with automatic revelation and tracking of experts' competencies. Such systems involve *information retrieval methods*.

There exist the following elements of an expert finding system (see Figs. 4.2 and 4.3):

- *Methods of competence definition.* In this context, two common methods may be mentioned: (1) expert competence profiles are created and then used for expert search, e.g., by information retrieval methods; (2) preliminary search is performed over given sources of information, and then the evidence of competencies is analyzed.
- *Sources of information and their models.* Possible sources of information are (1) artefacts associated with professional activity and related to a specific subject (e.g., project reports); (2) social and organizational sources providing information on the behavior of experts within their professional activity (including access patterns to information/interaction patterns within a project, information on events' attendance, as well as any other information used for expert connections).
- *Query models and information retrieval methods.* Query models specify definition and redefinition procedures for necessary competencies. Information retrieval methods (text methods, factual methods, etc.) determine the relevance of information (evidence) from different sources by an appropriate query.
- *Competence models.* They define the combination of evidence and perform weighting of sources for further assessment of experts' competencies.
- *Ranking methods.* Such methods are responsible for ordering of experts.

Classification of models. By analyzing different elements of an expert finding system, one can classify existing models by the following bases.

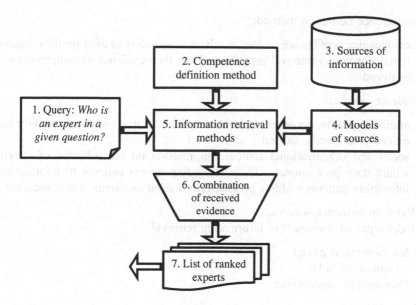

Fig. 4.2 The conceptual model of expert finding

Fig. 4.3 A typical diagram of information retrieval

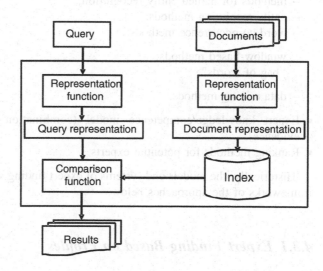

- Competence query modeling:

  - Topic-based Vector Space Model, text search, Latent semantic indexing, Boolean search;
  - Extended Boolean model;
  - query expansion including relevance feedback.

- Competence definition method:

  - competence profiles are created in advance, search runs after profile creation;
  - first, information retrieval is performed and then evidence of competencies is analyzed.

- Evidence sources:

  - artefacts (profile, documents, publications) associated with work activity and related to a specific subject;
  - social and organizational sources, information on the behavior of experts within their professional activity (including access patterns to information/ interaction patterns within a project, information on events' attendance, etc.).

- Models of evidence sources.
- Model types of document in information retrieval:

  - Set-theoretical models;
  - Algebraic models;
  - Probabilistic models, etc.

- Text analysis:

  - methods for named entity recognition;
  - entity resolution methods;
  - word co-occurrence methods:

    window-based methods;
    "bag of words";

  - data mining methods.

- Expert knowledge/competence model (combination methods for obtained evidence).
- Ranking methods for potential experts.

  Historically, the models and systems of expert finding were designed within the frameworks of the approaches below.

4.3.1 Expert Finding Based on Profiles

According to the initial approach [28, 68], all documents associated with a candidate expert make his *personal profile*. A document is assigned to a candidate in the case his authenticated authorship or if candidate's name (or his e-mail address) is mentioned in the text. A query with specified competencies serves for retrieving the relevant profiles of candidates (treated as common documents), and a user obtains a list of ranked candidates.

Later on, researchers suggested more sophisticated approaches employing state-of-the-art technologies in text information retrieval. The paper [108] settled the expert finding problem with required competencies via (1) constructing profiles with all organizational documents and (2) applying *latent semantic indexing methods*. Other investigators involve *language modeling* for information retrieval and rank candidate experts by the probability of query generation by the language model of candidate's profile [13].

4.3.2 The Document-Oriented Approach

Within the *document model*, the process of expert finding can be described as follows:

(1) by an appropriate query, select a collection of documents;
(2) for each document in this collection, establish persons connected with the document (by assumption, such subjects possess necessary knowledge in a corresponding subject domain).

This approach may also engage language models (see Model 2 in [13]).

4.3.3 The Window-Based Approach

Here, relevance assessment covers only a fragment of a given document (a text *window*) mentioning a potential expert. A text window may have fixed [69] or variable [15] size.

4.3.4 The Approach Based on User's Click-Through

Click-through data (a sequence of user's clicks) on earlier queries may provide additional evidence during new queries. In other words, it is possible to rank potential experts depending on the choice of documents associated with them [70].

4.3.5 Query Expansion

Consider query modeling. A query expresses the user's need for some information and, generally, represents a sequence of few key words. It is necessary to define an appropriate level of query representation detail and a method of user's query "enrichment." One can obtain a detailed description of desired information by *query expansion.* An initial query is enriched by choosing terms from documents relevant to this query. Particularly, it is possible to consider two conditions:

- without user *feedback*, high-rank documents resulting from query response are regarded as relevant [95];
- a user chooses few documents [16] relevant to a query in his view (these documents and the query are utilized in further search).

4.3.6 Expert Search Based on Different Networks

There exist the following classification bases for the models of *expert search on graphs*.

- Revelation of social relations:

  - formal relations (defined explicitly in user's profile);
  - informal relations (defined by communications of users):

    personal means of communication:

    e-mail (private messages, "one-to-one" relations) or private messages in online social networks and instant messaging services (sender-recipient analysis);
    phone call logs (caller-listener analysis);

    virtual communities (professional communities of practice and general communities of interest):

    posting services ("one-to-all" relations): analysis of message senders and headers;
    forums (collective blogs): message flow analysis including its structure;

  - similarity (relations are revealed, e.g., by the similarity of interests of correspondents, demographic characteristics or even location);
  - co-occurrence (relations are revealed by co-authorship in publications, projects, etc.).

- Analysis of relations among different subjects:

  - personal multirelations (e.g., in online social networks) and professional multirelations (e.g., in correspondence);
  - strong and weak relations;
  - relation strength assessment:

    objective (by analysis of documents and the structure of users' interaction);
    subjective (by users' appraisals):

    direct assessments;
    trust assessments and reputation assessments;

recommendations.

- Analysis of social networks (revelation of relevant nodes or groups and their peripheral counterparts):

  - centrality indices

    structural centrality indices for closeness or betweenness;
    link analysis measures: PageRank and its modifications (random walk models for different environments and virtual communities), HITS (Hyperlink-Induced Topic Search) and its modifications;

  - "expert" indicators:

    pairwise score of nodes by the level of expert knowledge;
    assessment by the "upstream-downstream" principle (the simplest method of relevance estimation is by the number of subjects with a smaller level of expert knowledge).

In the expert finding problem, it is useful to consider *social relations*. One can reveal the existing relations of candidates (potential experts) based on their co-occurrence in documents (i.e., a document represents a context merely). The matter concerns e-mail messages or personal posts of users in online social networks. Generally, such messages contain sender/recipient fields which assist in reproduction of social networks [100, 106].

According to a natural assumption, the status of a network member (in organizational or social networks) predetermines his importance for other members. For instance, in a narrow-subject community, an authoritative member can "answer" all popular questions on the subject. It is possible to involve *link analysis methods* [113] to study the importance of network nodes. Link analysis appears widespread in traditional field of information retrieval. We mention *PageRank* [83], an algorithm used in Google search engine, and Kleinberg's *HITS* [57], a link analysis algorithm for obtaining hubs and authorities in World Wide Web.

Following these considerations, Campbell et al. [27] compared the document-oriented approach with the approach based on link analysis (*HITS* algorithm). A system of e-mail messages within a certain organization was taken to construct a directed social graph using message headers and sender/recipient fields. It turned out that, in comparison with the document-oriented approach, *HITS* algorithm with *authority* estimates guarantees higher accuracy of candidates' ranking, but demonstrates lower efficiency (the size of a corresponding network is small).

Revelation of ties in posting services seems more difficult. A post has a single sender (a unique sender field), whereas the recipients are all members of a virtual community. Information exchange in (professional or common) forums takes place in discussion threads. The paper [118] studied a large narrow-subject community of experts in *Java* programming questions. A *social graph* was constructed by post/reply analysis of users' interaction; the arcs of such graphs have directions from questions to answers. An arc means that a respondent possesses more expert knowledge than an enquirer. A network defined by such a graph can be called a *competence network* (or competence graph). The authors adopted the following metrics for potential experts' ranking: the ratio of questions and answers, *HITS* and *PageRank* (actually, *HITS* and *PageRank* yielded best results).

Expert finding in [65, 104] has similar organization. Two approaches to competence graph construction are separated out. According to the first approach, competence graph construction employs query-relevant subject messages and the authors of such messages. The second approach considers the subject flow of messages (discussion thread). The paper [102] involves message flow analysis for identifying the reputation of discussion participants.

The above-mentioned approaches take into account only documents (artefacts) as the evidence of competence. Moreover, they presuppose that such documents necessarily contain (a) *key words* connected with required competencies and (b) named entities related to an expert. In other words, search focuses on documents belonging to a given subject, with subsequent establishment of its author or experts cited.

Clearly, in practice search may have another scenario. In particular, it is possible to find a relevant document; to identify persons mentioned in this document; to obtain documents recommended by these persons, i.e., "authorship by recommendation"; and finally, all these documents are used to find a "real" expert.

The chain "document"-"person"-...-"document" in competence search can be modeled by infinite random walk and absorbing random walk [103]. Such models serve for computing "most probable" experts.

4.3.7 Recommendation Models

A multiagent system, where agents can make and follow recommendations is called a *referral system* [117]. Here some agents cooperate and help another agent to find relevant information, so long as each agent possesses certain expert knowledge and knows his neighbors (information on their knowledge can be revealed, e.g., by the vector space model). An agent adopts the following procedure of functioning. An incoming query is assessed to match knowledge. If such match is established with a given confidence level, the agent sends back his answer; otherwise, he resends the query to neighbors with sufficient relevance to the query (according to his viewpoint).

It seems promising to apply these models with *pyramid incentive schemes*, when payments to experts grow with the number of other experts involved by him. Such incentive schemes are widespread in Internet projects, where audience expansion takes place owing to *virus marketing*.

4.4 Information Systems of Expert Finding

Nowadays, there exist numerous *information systems of expert finding*. Actually, they form two groups, namely, *corporate* and *online systems* of expert finding.

4.4.1 Corporate Systems of Expert Finding

Expert finding systems (especially, the ones used in geographically distributed organizations) provide capabilities to retrieve expert knowledge and related information (published documents, messages and other artefacts). Corporate systems of expert finding can be based on knowledge management services, e.g., documentation systems, community support systems, and information retrieval services.

Let us mention the following systems: *AskMe* (for *Microsoft SharePoint*), *Autonomy Universal Search*, *Tacit Knowledge* (embedded in *Beehive* developed by *Oracle*), *Recommind Mindserver*, *IBM Lotus Connections*, *SmallBlue*, and *TriviumSoft*.

Consider a corporate system called *SAP Expert Finder* [98]. For each employee of an organization, this expert finding system serves for expert finding by user profiles or different text sources (vacancy announcements, professional skills). For profile creation, a user is included in one or several communities (depending on tasks and requirements). A certain template of profile creation is suggested for all members of a community (description of skills, experience, etc.). The manager of an employee verifies the profile created by the latter, and then the profile appears in a search database. Expert profiles can be found through key words, Boolean operators and other search methods. Different groups of users have dedicated search scenarios. For instance, administrative purposes may require simple search by a user name in order to get the phone number and postal address of an employee. A researcher being in need of expert knowledge may require a search procedure yielding the competencies of experts (instead of their personal information).

SAP Expert Finder implements the following approach. On the one hand, expert profiles are created by users during self-appraisal. On the other hand, profile search adopts additional indexing of different text objects. The evidence of competencies consists in various artefacts (selected text objects, e.g., vacancy announcements). The system does not support any formal model of competencies; all matches or relevant text elements found in a database are used for expert choice.

4.4.2 Internet as a Source of Information on Experts

Common expert finding systems are generally intended for usage within an organization. Of course, there exist external experts and information on them may appear on public web resources (web pages, online libraries, blogs, etc.). The relevance of such information depends on its source; anyway, it allows assessing a potential expert and trusting him somehow. One can retrieve information

(1) throughout Internet (or its subset);
(2) within a certain online resource.

4.4.3 Expert Finding by Global Search Systems

In the case of known information sources, it is possible to acquire information and specify management rules for information flows using special "aggregators." *Yahoo! Pipes* (pipes.yahoo.com/pipes) represents a typical example of such systems with a user interface.

At the same time, expert finding may involve general- or special-purpose search engines (with some constraints, including automatic search by application programming interface (API) of search engines).

Information acquisition by search engines. Experts (or the evidence of competencies) can be found in different search engines such as *Yandex, Google, Yahoo!* and others. For instance, for evidence search, specify a query with the name of a potential expert, key competencies, the name of an organization, and certain constraints on source resources. The number of resulting documents makes a rough estimation of competencies.

Acquisition of information on intellectual clubs, networked expert and professional communities which reflect their activity on portals and websites. The latter provide registers of experts, agendas of seminars, materials of panel discussions, press releases, memoranda and resolutions of conferences, lists of speakers, and brief outline reports. These sources give valuable information on the competencies of experts and expert organizations (see Sect. 4.4.4).

Acquisition of information on competencies by news search. Well-known authoritative experts are often mentioned (cited, interviewed, etc.) in news of certain events. Among popular news services, we acknowledge Google News (news.google.com). News services perform automatic processing and systematization of news from very many sources; moreover, for each user they allow subscribing to personal interest news.

Acquisition of information on competencies by blog search. Blogs represent a powerful source of knowledge on personal competencies. Most blogs (including corporate ones) are indexed by search engines. For instance, we refer to *Google Blogs* (blogsearch.google.com).

Acquisition of information on competencies by scientific publication search systems. Scientific works can be found using *Google Scholar* (scholar.google.com) or *Microsoft Academic Search* (academic.research.microsoft.com). *Google Scholar* enables searching for basic publications in different disciplines and different sources. The rating of a publication is defined by its text, author, publisher and citation index.

4.4.4 Expert Finding Using Online Community Services

As a matter of fact, communities can be divided in two groups, namely, communities of interest and communities of practice.

A *community of interest* (COI) comprises subjects organized on a specific topic for exchange of information or participation in a joint interest activity.

A *community of practice* resembles a COI, but possesses the following distinctions from the latter:

- problems and tasks belong to a same domain/specialization (a COI deals with a common problem for different domains);
- new knowledge in a specific system of knowledge results from practice (in a COI, new knowledge results from synthesis of other knowledge from various domains);
- "formal" knowledge from a single domain (in a COI, knowledge embraces the opinions of all participants by reaching common understanding);
- group thinking with its inherent conformity;
- jointly used ontology (in contrast to the joint creativity and diversity of viewpoints in a COI);
- community members are rookies and veterans (a COI unites any interested persons from different fields);
- peripherical training (via informational participation). A COI includes persons with different interests/skills and has fewer restrictions; still, it is difficult to create common knowledge.

Among services of communities, we mention forums, collective blogs, mailing lists, groups, electronic brainstorming (see Sect. 3.5 of this book), etc.

Services of communities of practice. An example of such services is *ProfNet* [93], an online community of specialists in communication, which provides access to expert sources to journalists. Newspersons and other professionals (literary men, publishers, government employees, researchers) can contact experts and reporters to get answers to their questions. Such services imply competence profile creation during registration procedure (without restrictions, a user specifies key skills or knowledge domains, professional achievements, research fields, language qualifications and contact information). User queries are compared with existing profiles of experts. Dedicated employees may trace information on registered experts.

Services of communities of interest. For instance, *Google Answers* [8] represents a knowledge market (presently, this service does not support new queries' creation). For reasonable fees, users publish their questions and announce the expected remunerations for answers. Experts select questions based on their knowledge and announced remunerations. In this case, the asker gains an access to expert's knowledge as the result of negotiations. If an expert does not block a question (i.e., chooses an answer), users possibly have to raise the remuneration. Users can assess expert's knowledge only by his answers to their questions or the questions put by other users. In addition, users can register their assessments in the system.

A popular service called *Google Answers: Frequently Asked Questions* [82] allows a user to put a question for a whole society. A user can add tags (i.e., specify the category or key words of a question) to help potential experts find the question. As a reward for answering, potential experts receive rating points from a user's bonus fund. Actually, answers are provided by registered users based on their knowledge, experience and skills. All rating points go to an expert with the best answer. Users vote for a certain question and answer by leaving positive or negative assessments (the so-called social filtering), which increases or decreases their rating points. Users receive assessments of two types, rating points and reputation. The former is mostly increased by answering and participation in voting. The latter reflects user's contribution to community development—it grows with the number of high-quality answers.

There exist similar Q&A services (twitter.com, answers.yahoo.com) which provide the following capabilities: user registers, ready-made subject domain classifications (taxonomies) or the feasibility of their manual creation, expert profiles (e.g., text description) and feedback (generally, based on the assessments of community users).

4.4.5 Expert Finding Problem in Web 2.0

Today, many online web services possess characteristic properties leading to the concept of *Web 2.0* [92]. The major principle of *Web 2.0* concerns users' involvement in filling and multiple verification of content. Furthermore, online resources (or services) have become socially-oriented, services presuppose explicit representation of user profiles with social relations among them (social networks). Therefore, *Web 2.0* is remarkable for that the primary sources of knowledge (including expert knowledge) consist in users themselves. Explicit introduction of online user profiles (with a global identifier in the World Wide Web, e.g., the URL of profile's page) eliminates many problems connected with knowledge maintenance, knowledge integration and reliable knowledge retrieval. Explicit social relations are important for efficient processing of user data by a social network (here we mean joint creation and modification, structuring, retrieval, assessment and usage of professional knowledge) under deficient knowledge and incompetence.

Let us consider services in the cognitive aspects of expert knowledge creation (and other knowledge processes). Actually, services support the context of knowledge creation, thus facilitating the flow of ideas and conceptions, knowledge transition from one form to another. It is possible to identify the following groups of services:

- recording services, i.e., records of a user (an expert) added in reverse chronology during professional activity (they promote structuring and presentation of ideas, delayed self-communication, open debates and discussions in readers' comments);
- services of virtual seminars, electronic brainstorming, networked strategic conversations, conferences, etc. intended for discussions;
- supporting services for rational thinking (mind maps, cognitive maps and conceptual maps at the individual level; e-brainstorming, joint work on mind maps, cognitive maps and conceptual maps at the group level).

The first group comprises blogs, the second group corresponds to discussion forums, whereas the third group represents Wikipedia and online mind mapping services (www.mindmeister.com, bubbl.us), see Fig. 4.4. It seems that online social networks provide light support for professional knowledge creation; they aim at disseminating knowledge or information and at maintaining personal relations (instead of professional ones).

4.4.6 General Representation of a Socially-Oriented Service

Socialization has become possible owing to *Web 2.0* technologies. In the context of knowledge management, we reveal the general functionality of a socially-oriented service which provides special tools of knowledge operation (see Fig. 4.5). This is the capability of knowledge creation and integration: search system realization based on tags and microformats, the capability of free creation for the emergent structure of links to useful information sources, free structuring of content by tags, automatic content recommendation based on user history and profiles, and notification of new events.

The model of such a service incorporates three spaces, viz., social network, informational network and meta-informational network. Consider a typical user scenario as follows. A user creates and modifies content according to his interests (content tagging) on the occasion of a certain event. This content is commented and assessed by other users in the course of their communication and discussions. Imagine that the content appears interesting for a user; he may add content's URL to bookmarks (for further usage by a whole community) and the author to contacts (for news posting and future interaction if necessary). With the lapse of time, a social network (community) of interest may form with working, problem-related or business groups (long-term relations improve trust). In other words, social self-organization appears (see Sect. 3.8).

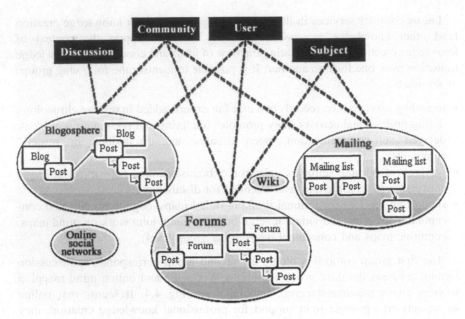

Fig. 4.4 Web 1.0 and Web 2.0

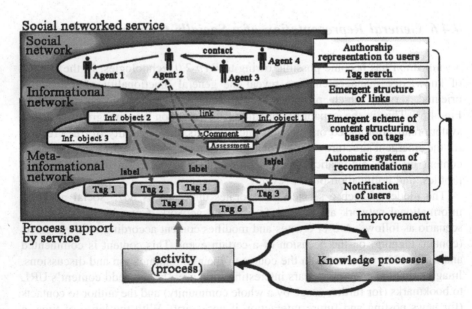

Fig. 4.5 A social networked service in the context of activity

4.4.7 Socially-Oriented Services Within or Beyond an Organization

The application of socially-oriented services in an organization can be considered in two planes (see Fig. 4.6):

(1) application for employees of a specific organization (internal communities). Employees enjoy the opportunity to fulfill their own creative potential via joint work (possibly, in project groups) without temporal or spatial restrictions using *Web 2.0*. Social relations of employees promote rapid creation and dissemination of knowledge via accelerated exchange and reutilization, as well as knowledge assessment including fast and efficient feedback. In this case, knowledge flows have any directions (horizontal, vertical and diagonal). In the final analysis, knowledge base gets accumulated, which is crucial for further activity (recall community feeling development and cultural evolution);
(2) application for potential partners, customers and other persons in order to interest and engage, establish relations, extent the primary audience through integration with web services.

In what follows, we outline possible usage areas of social services for internal communities:

- blogs of employees/project managers for discussion of current ideas, issues and problems (collective usage), as well as for self-communication and knowledge transformation from implicit to explicit representations (individual usage);
- wiki-service of knowledge exchange, collaboration, control and balance by participants (joint settlement of current project tasks, creation of online encyclopedias and knowledge bases within an organization, publications, brainstorming implementation);
- bibliography maintenance services (e.g., *CiteULike*) for joint management of references and citations, export/import of entries, their tagging, and adding papers from publisher's portals;
- certain services for collective accumulation of web links and their ranking (e.g., *Delicious*) in order to reduce information retrieval/selection efforts;
- mind support services (e.g., *Mindmeister*) for individual/joint work with mind maps using web sources, etc.

4.4.8 Socially- and Semantically-oriented Services in the Context of Expert Finding

Collaborative tagging systems (CTSs). Online resources available in *Web 2.0* allow ordinary users to add content to a system. In such resources, information (more specifically, informational objects) gets categorized through tags (labels or

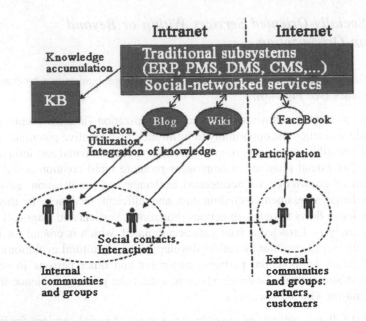

Fig. 4.6 Usage of services in internal and external communities of an organization

key words) arbitrarily selected by users. It is also possible to adopt existing tags (the term "folksonomy" corresponds to the "public" classification of information). Another option concerns user profile creation on the basis of tags involved by a user to describe new informational objects. These profiles may serve for designing recommendation systems [32] that suggest informational objects (e.g., publications) and users for potential collaboration, or follow the state of communities. Finally and obviously, to find an expert with desired competencies in such systems, one should merely create a profile these competencies, add appropriate tags and adhere to system's recommendations.

Social bookmark services. Actually, social bookmark services (*Delicious* (delicious.com), *StumbleUpon* (stumbleupon.com)) enable users to create personal bookmarks of valuable web resources. By default, all saved bookmarks of a user are available for public view. A user can systematize his bookmarks by assigning one or several tags. Collections of bookmarks reflect interests of users, and their publicity unites users in groups of interests. Generally, bookmarks are available for public view and categorized by tags, which facilitates rapid expert finding.

We refer to *Dogear* [72] (see Fig. 4.7) as an example of social bookmark systems intended for organizations. Numerous surveys testify that the system improves user's awareness of the knowledge and interests of other employees in an organization.

Blogs. There exist personal, collective and corporate blogs. A blog allows an expert to add records during his professional activity. And readers may comment his messages. Modern blogging resources (e.g., *LiveJournal*) support explicit

Fig. 4.7 The graph of
connections among users and
subject domains (based on
tagging in Dogear) [72]

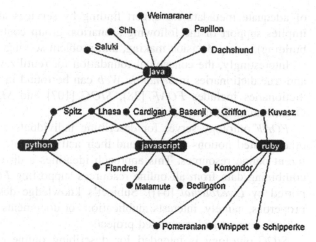

creation of social relations. And so, communities form around experts. Therefore, an expert can be found by social relations or any of the above search systems.

Online professional social networks. Any users registered in modern professional networks (http://www.linkedin.com, http://forums.e-democracy.org and others) are able to create and maintain lists of business contacts.

Such networks make it possible to maintain and expand user relations, to find companies, individuals and groups of interest, to publish CVs and seek for employments, to give or receive recommendations, to publish vacancies, and to create groups of interest. For instance, *LinkedIn* admits expert search by the following categories: competencies, location, (previous or current) affiliations, or acquaintance.

4.5   Expert Finding in Semantic Web

Semantic Web (or *Web 3.0*) is the next generation of Web, which aims at representing information in forms suitable for automatic processing and human comprehension. Standard Web provides information in the form of web page texts; users extract such information by web browsers. *Semantic Web* presupposes information recording as a semantic network with ontologies. Faster and better understanding among networked users is guaranteed by semantic interoperability standards and methods.

An initiative of *ExpertFinder* [35] endeavors to design a *Semantic Web* infrastructure with expert finding in Internet. The list of associated tasks includes the development of dictionaries and rules (e.g., for ontologies Friend Of A Friend (FOAF) and Semantically-Interlinked OnLine Communities (SIOC)), recommendations on annotation of personal and corporate webpages, conferences, publication databases and other web resources indicating of expert knowledge; provision of adequate metadata for expert finding by services and agents. The initiative implies support of the following scenarios: group control, group creation (team building), group decision-making, and problem solving.

Interestingly, the conceptual foundation for reutilization and binding of tried-and-true dictionaries in *Semantic Web* can be found in [21]. The list of possible dictionaries includes *FOAF* [38], *SIOC* [107] and *SKOS* (Simple Knowledge Organization System) [114].

FOAF ontology serves for describing individuals, groups, organizations and other related notions (subjects and their activities, interaction processes) on different online resources. This approach identifies a distributed social network by combining data from all online resources supporting *FOAF* (e.g., *FOAF* is supported by *LiveJournal* [64]). Subject's knowledge description engages special properties, namely, interests, publications or documents associated with a certain subject, current and completed projects.

SIOC ontology is intended for describing online communities (users, their messages and discussion sites). Thus, we identify a distributed online community by combining data from all online resources supporting *SIOC*. Particularly, it is possible to exchange information among resources. Expert finding in such communities employs some properties provided by *SIOC*. For instance, the property "topic" (sioc:topic) describes a category of a message. By uniting all topics of user's messages on all online resources, we easily discover the interests and knowledge of this user. Forums and sites (sioc:forums, sioc:sites) may have topics, and a user concerned with a topic becomes a subscriber (sioc:subscriber_of).

Let us make a remark on *SKOS* ontology. Actually, it can be applied for detailed description of topics (thesauri, classification schemes, taxonomies).

No doubt, implementation of the above ontologies requires definite efforts by online resource developers (possibly, by authors of informational objects). Meanwhile, *FOAF*-profile generation may involve user data from common *Web 2.0* systems. For this, take advantage of the approach [32], i.e., profile creation based on information from blogs or online publication databases (in fact, such profiles are used to supplement *ExpertFOAF* files, an extension of *FOAF* [47]).

In *Find an eXpert via Rules and Taxonomies* (FindXpRt) [63], *FOAF* ontology is augmented by new properties for explicit reflection of expert knowledge: expertise, rating, work duration. Moreover, certain rules from *RuleML* language [97] assist in deductive derivation of new facts on an expert. Such system can be used for expert finding and collaboration support.

Chapter 5
Active Expertise

In the preceding chapters, we have described e-expertise by assuming (implicitly or, par excellence, explicitly) that all participants of expert activity follow the primary goals of expertise, i.e., to provide maximum complete, adequate and reliable information on the object or subject of expertise for Principal's decision-making.

In particular, we have supposed that (a) expertise organizers form an expert group which is able to give most complete and objective assessments for the object or subject of expertise, (b) a methodological group chooses certain procedures of expert data acquisition and processing for generating the most compete and exhaustive answer and (c) experts proper report their true assessments, being guided purely by their professional knowledge and ethics. It is important that experts comprehend the goals and interests of a Principal. However, all participants of expert activity represent people with personal interests. Generally, each participant can

- demonstrate *strategic behavior* according to his personal interests (preferences, partiality);
- recognize the incompleteness of his *awareness* (about the goals, interests and opinions of other participants, various information);
- act *unconscientiously*, i.e., deceive or neglect professional duties.

In other words, participants of expertise can be active subjects, *rational* in the following sense. Under available opportunities, they strive for satisfying personal interests (including selfish ones).

In the context of the expertise organization problem, the major importance belongs to the case when some participants of expertise fail to be *"independentexperts."* An expert, several or even all experts can have certain preferences regarding the results of expertise, possibly contradicting the interests of a Principal or other experts. Such subjects can show *active (strategic) behavior*, i.e., exert certain impact on the process of expertise in order to distort it (thereby, making closer to a desired result).

Active expertise is a type of expertise, where the strategic behavior (biased actions) of some participants and their manipulation capabilities play an appreciable role. (Table 5.1)

D. Gubanov et al., *E-Expertise: Modern Collective Intelligence*,
Studies in Computational Intelligence 558, DOI: 10.1007/978-3-319-06770-4_5,
© Springer International Publishing Switzerland 2014

Table 5.1 Basic problems of manipulation

Object	Direction of Manipulation		
	Principal → Expert	Expert → Principal	Expert → Expert
Staff	1. Forming a desired staff of an expert group	6. Making a decision to participate (or refuse from participation) in an expert group by an expert	9. Influencing the decisions of other experts to participate in an expert group
Interests	2. Applying a personalized incentive scheme (for each expert) or a unified motivation scheme (for all members of an expert group)	7. Offering financial guarantees (compensations, honoraria) for obtaining a desired expertise result for an expert	10. "Income" sharing with other experts for obtaining a desired expertise result
Set of feasible messages	3. Forming a query to experts or a scenario and conditions of expertise conduct (see Sects. 3.2 and 3.3)	–	–
Expertise procedure	4. Choosing an expertise procedure (e.g., processing methods for experts' opinions) for obtaining a desired result	8. Misrepresenting information reported by experts	11. Coordinating reported information with other experts
Awareness	5. Forming certain beliefs of experts about the opinions of other members of an expert group	–	12. Forming certain beliefs of experts about the opinions of other members of an expert group

Expertise procedures with neutralized manipulation capabilities (more specifically, where all participants benefit by truth-telling and conscientious actions) are called *strategy-proof expertise procedures*. For detailed discussion of strategy-proof expertise, we refer to [26, 54, 55, 56, 80].

Goals of manipulation. We have mentioned that manipulation in active expertise lies in strategic actions of participants in order to modify the result of expertise according to their goals. Which are these goals? There exist three general groups of *goals of manipulation* by experts or coordinator:

(1) to achieve a personal desired result of expertise or to make the final result of expertise as close to this result as possible;

(2) to avoid a personal undesired result or to make the final result of expertise as far from this result as possible;

(3) to increase the personal influence (rating, reputation) of an expert as much as possible, even with prejudice to Principal's position.

The last situation appears widespread among experts with regular participation in expertise. In the last analysis, it facilitates satisfaction of goals 1 or 2 (in addition to personal ambitions of an expert).

Subjects and objects of manipulation. Clearly, participants of expert activity have different motivation to manipulation, different capabilities and methods of manipulation. To systematize them, we should identify *subjects of manipulation* (who performs manipulation) and *objects of manipulation* (what is manipulated).

For instance, a coordinator can influence the result of expertise by forming the *staff* of an expert group or an *agenda* (a list of discussed questions and possible answers to them). Moreover, he can suggest a certain procedure of e-expertise for "pushing through" a required expert decision. On the other hand, experts can distort reported information, thus hyperbolizing the advantages of an expertise result suggested by them (striving for personal *professional interests* or *lobbying* the interests of some third parties).

It is possible to classify experts as follows: experts working on Principal's order and honorarium; experts representing the interests of lobby structures (e.g., business companies) and being paid by such structures; experts representing the interests of public organizations, clubs, networked communities, etc. Experts may have no information on their clients and receive their honoraria from an "independent" fund.

Therefore, it seems reasonable to classify different manipulation phenomena in expertise based on some formal model of active expertise. Such approach assists in comprehension of nonformalized (qualitative, implicit) components of manipulation.

Most adequate formalization frameworks for manipulation processes include decision theory, game theory [76], collective choice theory [1, 9], and theory of control in organizations [26, 80]. A common feature of these frameworks is treating expertise as a *game*, i.e., a mathematical model of rational subjects' interaction.

According to the assumptions on expertise structure [see Fig. 2 (The interaction of basic participants) of Inroduction], the game-theoretic model of expertise is described by the following components:

1. *The staff of expertise participants*:

   1.1. a Principal[1] (P);
   1.2. experts (E);
   1.3. a moderator (M).

2. *The goals and interests* of expertise participants in terms of correlation between desired and implementable results of expertise.

---

[1] In this section, the conception of a Principal covers a Principal proper (who makes decisions based on expert information), the coordinator of expertise or third parties interested in expertise results.

3. *The mechanism of expertise*, which consists of:

   3.1. the sets of feasible messages (answers, assessments) of experts;

   3.2. an expertise procedure (see Chaps. 2 and 3), including

      3.2.1. the number of interaction periods of experts and the period of expertise;

      3.2.2. the order of reporting by experts;

      3.2.3. the methods of expert information acquisition and processing to obtain expertise results;

4. *The awareness* of all expertise participants about:

   4.1. the object or subject of expertise;

   4.2. other components of the model (the order of reporting, expertise procedure, etc.);

   4.3. the awareness of other participants;

   4.4. the Situation awareness [118].

Each of the stated components in the expertise model (staff, goals, mechanism and awareness) can be the object of manipulation.

In principle, any expertise participant may perform manipulation. Complete description calls for considering the exhaustive list of persons concerned, *viz.*, direct participants of expertise and other persons being able to influence expertise participants. However, this chapter focuses on the problem of strategy-proofness in expertise. Therefore, we believe that the list of expertise participants comprises just two persons–a Principal (simultaneously acting as a moderator) and experts.

Consequently, such assumption defines three *directions of manipulation*:

1. a Principal influences experts;
2. experts influence a Principal;
3. experts influence experts.

The problems of expertise results' manipulation: a classification. By systematizing the objects and directions of manipulation, one easily outlines 15 possible basic problems of manipulation (see Table 5.1).

Yet, situations when experts directly influence the sets of feasible messages or the a priori awareness of a Principal could be hardly imagined. And so, there are 12 problems of manipulation with serial numbering in Table 5.1. In the sequel, we analyze them in brief and overview the existing research results of corresponding formal models.

5.1 Principal Influences Experts

The following questions may arise naturally. Why should a Principal or coordinator of expertise manipulate experts? Would an appropriate term be *control*? In this context, we outline several key features. First, any expertise is conducted under existing external conditions (normative and legal, informational, techno-logical and other conditions); they can impose essential constraints on expertise mechanisms and apply certain requirements to expertise results. Second, a Principal can lie on an intermediate level in the hierarchical management system of a large organization. By manipulating expertise results, he influences decision-making at higher levels of the hierarchy. Third, a Principal may conceal his actual intentions. And fourth, in some situations a Principal appears unable to express his informational needs and intentions. Therefore, any managerial actions of a Principal, directed towards achievement of other goals (differing from officially declared ones during expertise) represent *manipulation*.

To proceed, consider Principal's capabilities of influencing different objects of manipulation (see Error! Reference source not found.).

1. Principal manipulates the staff of experts. Actually, guaranteeing a required result of expertise by forming a "specific" *staff of an expert group* makes a "classical" example of manipulation (especially, in normative expertise). Take a Principal demonstrating strategic behavior; from his viewpoint, the problem of expert staff formation admits the following statement. Find the minimal number of controlled experts to-be-added in the staff for ensuring a desired result.

Besides normative expertise, a similar type of manipulation is widespread in corporations' management (decision control at stockholders' meetings [7, 37]).

Formal analysis of such manipulation may involve theoretical results from staff control problems in organizational systems [79, 80].

This type of manipulation by a Principal can serve for mercenary ends (while defining the required number of "decoy" experts) or for establishing the presence of such manipulation in expert procedures and its consequences.

In political sciences there exists a series of research works focused on manipulation issues at elections [5] and voting in stock corporations [7, 37]. The cited publications place an emphasis on organizational and normative anti-manipulation measures. Other investigations are dedicated to the approaches of expert group formation with accelerated *convergent* decision-making, see [89].

In the case of e-expertise, this type of manipulation can attract Principal's attention during open expertise (the number and staff of participants is not fixed). Unfortunately, this issue has not still been analyzed.

2. Principal manipulates the interests of experts. Another classical type of manipulation lies in "purchasing" the opinion of an expert ("bribery"). Due to the variety of methods of such manipulation, the importance of revelation and anti-manipulation techniques in practice, we study merely most significant aspects of such manipulation in e-expertise.

This type of manipulation belongs to the class of *motivational control* [26, 80]. Suppose that it is possible to estimate (positive or negative) payoffs of all or specific experts from each possible result of expertise. Then it appears possible to estimate the "incentives" (or "penalties") to-be-assigned to experts by a Principal for obtaining a desired result of expertise. In other words, one can design an appropriate "incentive scheme" for experts.

Here we should discriminate between two cases as follows. In the first case, a common (*unified*) incentive scheme is constructed for all experts to motivate them report required data to a Principal. Solution of this problem may employ classical approaches from microeconomics [42] and theory of control in organizations [26, 80]. In the second case, a Principal applies *personalized* incentive schemes to specific experts (generally, in secret from other experts). This type of control actions can be interpreted as *corruption*. Mathematical models of corruption are explored in [45, 96, 110].

In the conditions of e-expertise (very many participating experts or an open expertise procedure), the given type of manipulation occurs when a Principal is able to choose most authoritative experts (whose opinions are considered by residual experts or whose assessments have highest weight). Meanwhile, e-expertise procedures generally possess openness, and the costs of such manipulation may grow significantly.

Principal manipulates expertise mechanism. In this book, an *expertise mechanism* is a set of the following components:

1. query to experts with questionnaire formalization (questions) and an order of their presentation to experts;
2. feasible answers to posed questions;
3. moderator choice and the influence of moderation definition;
4. an order of interaction among experts, the object or subject of expertise, a Principal during formation and reporting of expert assessments;
5. acquisition/processing methods for expert messages.

According to the game-theoretic approach, it seems reasonable to decompose this set in two parts:

1. A *set of feasible messages* (components 1 and 2 above).
2. *Expertise procedures* (components 3, 4 and 5 above).

3. Principal manipulates the set of feasible messages of experts. This type of manipulation possesses a broad range and includes, in the first place, achievements of weakly formalizable theories (psychology, linguistics, etc.). The form and sequence of questions, the list of feasible answers has a strong impact on the final result. Among classical examples of such manipulation, we should mention Zenon's aporia (e.g., *Achilles* and the *Tortoise*).

Today, e-expertise may involve certain approaches to formalize such manipulation and response to it (of course, after appropriate adaptation). We highlight investigations focused on analysis and formalization of *decision-making psychology*. Furthermore, the framework of *cognitive modeling* assists in verifying the correctness of posed questions and possible substitution of notions during elaboration of decisions [2]. Final decision may depend strongly on the precise statement of questions to experts.

According to *agenda theory* [17, 71], almost any decision can be guaranteed by choosing the content and sequence of questions (under existing information on the preference of all experts). This fact allows designing agendas to achieve necessary final decisions.

Modern economic research of decision-making provides more and more evidence of the following phenomenon. People often make decisions being guided by *herd instinct* [30] or previous experience [30, 34]. In particular, we refer to *situational control*, usage of *typical decisions*, etc. By a natural conjecture, such behavior damps the impact of manipulation. For e-expertise, a substantial role belongs to that experts can interact with each other on discussion sites.

Finally, there exist mathematical models showing the *conformism* of agents [41]. A promising direction of research concerns exploring the impact of conformity behavior on Principal's manipulation capabilities.

4. Principal manipulates the procedure of expert messages' treatment. A Principal may perform manipulation by choosing certain procedures of acquisition and processing of expert messages (including the order of reporting and the number of iterations), which lead to a desired result of expertise under other parameters fixed.

Such manipulation takes place in the distribution of seats among several political factions in a national parliament according to the proportional representation principle. By selecting a rule of seats' definition, one can make a certain faction uninfluential; in other words, if this faction joins any coalition, the latter fails to become a *winning* coalition (to push through a decision supported by this coalition). Influence level analysis for specific factions bases on *influence indices* [6, 75]. They can be used for manipulation (selection of a proportional representation rule) and for prevention of such manipulation.

Another modeling tool for manipulation processes and anti-manipulation measures is proposed by *implementability theory* [48]. This framework assesses the feasibility of implementing a specific aggregation procedure of expert appraisals under strategic behavior of experts.

Theory of organizational systems gives strategy-proof *consent mechanisms*[2] [26]. Some direction (e.g., of investing) is selected as the *basic* one (the development of other directions is impossible without the development of the basic direction). Expert commissions are created for other directions to generate a consentient decision on the shares of investments into the development of these directions with respect to the basic direction (the number of expert commissions equals the number of directions minus 1). The information reported by an expert commission is used to allocate available financial resources to a corresponding direction. Obviously, the final allocation of resources appreciably depends on the choice of the basic direction. This is the Principal's leverage during manipulation.

E-expertise models [43, 44] study the procedures of weighted assessments, where the weights of experts are determined using their voting history. Interestingly, the choice of weights' definition (rating assignment rules for experts) strongly affects the behavior of experts and expertise results.

---

[2] The term "consent mechanism" also means the mechanism of a group consensus which serves for elaborating consentient decisions on the goals and ways of their achievement.

5. Principal manipulates the awareness[3] of experts. Such manipulation represents *informational control* [78, 80]. There exist three types of informational control:

(a) influencing the beliefs of experts about a subject domain (*informational regulation*);
(b) influencing the beliefs of experts about the awareness of their colleagues (*reflexive control*);
(c) reporting some information on the expected result of expertise (*active forecasting*).

A model analyzing the capabilities of informational control of experts is examined in [78]. The authors demonstrate that, under a fixed treatment procedure of expert messages, a Principal can ensure almost any result of expertise by reporting confidentially appropriate information on the opinions of certain experts to other experts.

In comparison with conventional expertise, the framework of e-expertise provides considerably smaller manipulation capabilities for the awareness of different experts. A key notion in the theory of informational control is decision *stability* [78]. In terms of expertise, this notion means that the result observed by experts must coincide with their expected result (under the information on the beliefs of other experts reported by a Principal). If the Principal performs informational manipulation, the final decision appears unstable in most cases. Subsequently, experts may distrust the information reported by the Principal. However, stability is not so important in one-shot expertise. The interaction between experts and a Principal in e-expertise takes place many times, and the property of stability plays a major role, which restricts the capabilities of informational manipulation.

5.2 Experts Influence Principal

This direction of manipulation is traditional in *social choice theory* and *theory of organizational systems*. Nevertheless, the general interpretation of manipulation is somewhat wider. Considering different objects of manipulation, we dwell on most important results of these theories in the context of e-expertise.

6. Expert manipulates the staff of experts. Actually, the staff of an expert group is selected by a Principal; therefore, the only direct influence of an expert on the staff of an expert group lies in his rejection from participation in expertise. Suppose that the procedure of expert group formation involves experts (see formation methods for expert groups in Chaps. 2 and 4). In this case, such procedure can be treated as a component of a corresponding expertise mechanism, whereas

---

[3] The awareness of a subject is information on essential parameters and the awareness of other subjects, used by this subject in his decision-making.

experts' recommendations act as a part of information supplied by experts to the Principal. In the course of expertise, an expert can provoke discussions of different questions not related to the object or subject of expertise, thus "guiding" it away. And so, the object of manipulation is a specific implementation of expertise mechanism.

7. Expert manipulates the interests of Principal. Unfortunately, a priori there exists a possibility that an expert influences decisions of the Principal (e.g., by suggesting certain preferences to the latter and provoking his corruption behavior).

We identify two types of material motivation of the Principal by an expert: (1) classical corruption ("bribery" payment to the Principal, see references in Sect. 5. 2) and (2) when the rules of financial participation (or responsibility) of experts form a legitimate component of a decision-making procedure [42, 49]. The second type corresponds to manipulation of expertise mechanism.

Expert manipulates the set of feasible messages, thereby causing deviant processes in expert decision-making. We believe that this case of manipulation is inadmissible. Really, if an expert participates in formulation of questions for expertise, this is a component of the procedure of expert information treatment (acquisition and processing), and the set of feasible messages includes information an expert may report to the Principal during expert group formation.

8. Expert manipulates the procedure of expert messages' treatment. An expert manipulates a procedure in the following way. Under a procedure selected by the Principal and a well-defined set of feasible messages, an expert chooses a message leading to his maximal utility. This agrees with the *classical comprehension of "manipulation."*

We underline that most results of formal models' analysis in social choice theory and theory of control in organizations apply to exactly this case of manipulation. A major problem consists in designing strategy-proof mechanisms. The nontrivial character of this problem directly follows from a classical result of social choice theory known as *the Gibbard-Satterthwaite theorem* [9]. It claims that, under arbitrary preferences of experts, the only strategy-proof mechanisms are *dictator mechanisms*, where the final appraisal is dictated by an a priori defined expert (the opinions of other experts becomes inessential). Nevertheless, under certain assumptions, there exist other (not so trivial) strategy-proof mechanisms of expertise. Most research in social choice theory concentrates on derivation of existence conditions for nondictator strategy-proof mechanisms (e.g., see the overviews [19, 23, 74]).

Nowadays, the growing popularity belongs to the following approach. Possible losses from manipulation in common procedures of expert assessment aggregation are estimated by:

1. *the closeness of expertise results* in the cases when (a) experts perform manipulation and (b) experts adhere to truth-telling [66];
2. *the degree of manipulability* (the relative frequency of cases, where experts benefit from manipulation);
3. the maximal possible deviation due to *manipulation*.

For the time being, investigators have demonstrated that these losses are moderate for many mechanisms. Furthermore, in several cases experts loose more than gain by their strategic behavior. However, the following issue from implementability theory [49] remains open. Given a certain procedure enjoying optimality in the absence of manipulation, should one apply this procedure in the case of manipulation or construct another procedure? The answer exists only in special cases.

E-expertise adopts calculation mechanisms for the *reputation* of experts depending on the history of their participation in previous expertise. Here of crucial importance are the results on designing multicriterion strategy-proof mechanisms of active expertise [19]. Under some assumptions on experts' preferences, it is impossible to build strategy-proof mechanisms (i.e., each expert provides true information on each issue) such that the reputation of experts depends on voting history.

The above-mentioned research *par excellence* deals with models, where experts consider their personal interests. In this case, the effect of manipulation gets manifested in "hogging the cover," i.e., if an expert believes that the result of expertise would not coincide with a desired result, he reports *extremal assessments* (towards the desired result).

For e-expertise with reputation calculation based on the history of agents' participation in previous expertise, some researchers explore conform behavior, see [41]. Notably, they show that, in a series of cases, experts may strive for divining the final result of expertise and report it to improve their rating/reputation. It seems interesting to study models, where the strategy-proofness of mechanisms is guaranteed by counteracting two directions of manipulation–agents pursuing a certain result of expertise and agents maximizing their rating.

Expert manipulates the awareness of Principal. We believe that this case of manipulation is inadmissible; indeed, an expert directly influences a Principal only through information and communication channels used in expertise mechanisms. If an expert reports untrue information in response to Principal's queries, we obtain the type of manipulation described in case 8.

Expert influences other experts. A distinctive feature of e-expertise is that a Principal possesses limited capabilities to control the interaction of experts. Therefore, a topical direction of investigations concerns prevention methods for purposeful influence of certain experts on other experts. This type of manipulation attracts major interest in organization and conduct of e-expertise.

5.3 Experts Influence Experts

9. Expert manipulates the staff of experts. Under the above assumptions, we characterize this manipulation as an expert's influence on the process and result of decision-making of other agents (whether they participate in an expert group or not). Note that the capabilities of such manipulation are strongly limited. Really, if this influence gets regulated by an expertise mechanism or has the motivational or informational character, it belongs to other types of manipulation (see below). By analogy to manipulation focused on the staff of experts and performed by an expert or a Principal, here feasible actions are decisions to participate in expertise or not to participate. In other words, an expert can decide to participate in expertise to engage or exclude other experts. The result of such influence and corresponding decision-making process admits formal description by models of conformity behavior, models of opinions' propagation in social networks or virus models [50].

Suppose that experts may coordinate their decisions on participation in expert groups. Then modeling and analysis of their behavior can employ certain theoretical results from staff formation problems in organizational systems with coalitions, network formation games [50]. However, the experts' decisions on collaboration based on these models, to a large degree, depend on current expertise mechanisms. Therefore, such collaboration actually represents manipulation of the procedures of expert messages' treatment, which is performed by experts (see case 11).

10. Expert manipulates the interests of other experts. In this case of manipulation, an expert influences the messages of other experts by offering certain incentives for their services, concessions (queries). This phenomenon can be viewed as the process of coalitions' formation (*collaboration, virtual collaboration*), where a manipulating expert engages other experts into a coalition through payoff reallocation among other members of the coalition. Such collaboration of purposeful subjects has been intensively studied in *theory of cooperative games* with transferable utility [87]. The results of these investigations can be applied (a) to analyze the conditions of coalitions' formation by experts, (b) to find expertise results convenient for coalitions and (c) to evaluate the payments of a manipulating agent to other members of a coalition. Interestingly, most publications in this field get focused on the following issue. When is the *maximal coalition* (i.e., the one uniting all experts) formed? Appreciably less attention is paid to coalitions composed of some agents [22].

There exist generalizations of the described results to the case of networked interaction in e-expertise problems. For instance, we refer to a detailed survey in [50]. Moreover, another thriving direction of theoretical research covers resource allocation problems on networks [51]; the solutions of these problems can be used to compute the payments of a manipulating expert to other experts engaged in a coalition (with proper consideration of the specifics of networked interaction).

Expert manipulates the set of feasible messages. Similarly to the case when an expert manipulates a Principal, we believe that an expert is almost unable to influence the set of feasible messages of other experts.

11. Expert manipulates the procedure of expert messages' treatment. Here we assume that an expert manipulates his personal request (see case 8) and tries to coordinate such manipulation with other experts (e.g., by "calling" them to distort their messages). In contrast to the situation discussed in case 10 (influence presupposes motivation), here the only tool of influence consists in communication among experts (to coordinate their positions). To model such interaction, it is possible to involve certain results from theory of cooperative games with *non-*

transferable utility [84]. This framework serves for studying possible factions in political science [12, 75], where the feasibility of coalitions' formation from several parties is analyzed via their influence indices [6].

Among approaches to prevent such manipulation, we mention research in the field of designing *coalition strategy-proof* mechanisms of decision-making [18, 60, 109]. Here an emphasis is made on the conditions when strategy-proof mechanisms simultaneously become coalition strategy-proof (i.e., any set of experts never improves expertise result by distorting their true requests in a coordinated way).

In the sense of e-expertise, we associate particular interest with game-theoretic models of network formation [11, 20], which can be interpreted as a generalization of theory of cooperative *games with non-transferable utility* for networked problems. These investigations yield new concepts of an equilibrium to describe the coalitions of experts appearing in the course of active e-expertise and to analyze the stability of these coalitions against destructive influences.

12. Expert manipulates the awareness of other experts. This type of manipulation can be treated as "bluffing," i.e., an expert deliberately deceives other experts on his capabilities or awareness used in decision-making. Due to intensive communication among participants of e-expertise and Principal's limited capabilities to control such communication, this type of manipulation becomes one of most "dangerous."

Any actions of a manipulating expert actually represent *informational control*. Hence, it is possible to identify three basic types as above (see informational manipulation of experts by a Principal). Nowadays, there are few publications studying models of informational control applied to an agent (a participant of a decision-making process) in order to influence other agents (e.g., see [78]). Particularly, this is the case for decision-making by cognitive and/or game-theoretic modeling. At the same time, many authors explore the impact of such phenomena from the viewpoint of psychology and other sciences. Therefore, in-depth analysis of this type of manipulation represents a most promising direction in theory of e-expertise.

This chapter has overviewed the existing models of active e-expertise and manipulation capabilities for e-expertise results.

The suggested classification of methods and objects of manipulation serves to structure many publications in this field and to choose appropriate strategy-proof mechanisms of e-expertise.

The results described above might and should be considered as a "dual-purpose weapon." They can be adopted to prevent manipulation (for a good cause) and to organize manipulation (for a bad cause) in e-expertise. However, it seems obvious that, even in conventional expertise, the problem of active participants has not still been completely solved. The specifics of e-expertise bring some types of manipulation in the forefront (e.g., informational manipulation of other experts by a given expert) and compensate other types of manipulation (in the first place, Principal's influence on experts).

Conclusion

This book proposes e-expertise technologies which have been tested in practice, including the components of preliminary analysis of a problem situation and query preparation for experts, expert finding, acquisition of expert comments and assessments, construction of conceptual models and substantiation of draft decisions based on them. Generally, implementation of expert technologies possesses the corporate (relatively close) character. At the same time, the accumulated experience in the field of e-expertise organization and conduct can be found in different portals of networked expert communities.

The following research domains seem promising in further development of e-expertise technologies:

- self-organization phenomena in expert communities;
- motivation mechanisms for joint work of experts in networked expert communities;
- normative and legal conditions of e-expertise implementation and status determination of e-expertise results;
- formation procedures for networked expert groups settling applied problems;
- dynamic calculation of "reputation/rating" of experts;
- organization methods for hierarchical networked interaction of experts.

We make special mention of the development and testing of typical software tools of e-expertise, customizable to the specifics of applications.

Successful solution of the problems outlined in this book requires joint efforts of researchers (experts in control/management, mathematics, information security, psychology, sociology, etc.) and proper public attention.

D. Gubanov et al., *E-Expertise: Modern Collective Intelligence*,
Studies in Computational Intelligence 558, DOI: 10.1007/978-3-319-06770-4,
© Springer International Publishing Switzerland 2014

Appendix A
Supplementary Information and Typical Documents

A.1 Typical Statute of Networked Expert Community

1. Purpose.

1.1 This Statute defines the legal foundation, principles of organization and basic directions of formation and activity of a Networked expert community (hereinafter referred to as an Expert community).

1.2 The legal foundation of activity of a expert community is the legislature, normative and legal acts of a country, which regulate organization and conduct of networked expertise, as well as this Statute.

2. Goals and tasks of expert community.

2.1 The activity of an expert community aims at:

- improving the quality and standards of living;
- maintaining the stable development of the national economy;
- strengthening of national and public security;
- enhancing the competitive ability and international prestige of the country;
- increasing the efficiency of activity and prosperity of an expert community.

2.2 An expert community performs the following tasks:

- expert support of interaction between state authorities, business companies and citizens;
- permanent quality growth of managerial decision-making;
- detection and prevention of potential threats and risks;
- expertise of draft versions of normative and legal documents;
- formulation of key problems and ways to settle them;
- search for new, original, innovative capabilities;
- search and assessment of factors having a strong impact on the development of socioeconomic events;
- substantiation of managerial decisions depending on interests of different parties;

- support in formation and control of programs, projects and strategies of organizations;
- assessment of individual contributions by participants of decision-making;
- reduction of production/servicing costs.

3. **Membership in an expert community**.

3.1 Possible members of an expert community are the representatives of:

- federal, local and municipal authorities;
- independent expert communities;
- business companies;
- scientific and educational organizations;
- public organizations;
- political parties;
- civil society institutions (clubs, discussion panels, etc.);
- other professional and expert organizations, private experts.

3.2 Full members of an expert community are persons confirming their participation in this community via registration on the official website (portal) according to an established procedure of accreditation (certification).

4. **Principles of activity organization in expert community**.

4.1 The activity of an expert community bases on the principles of legitimacy, observance of the rights and liberties of individuals and citizens, the rights of a legal entity, as well as on the principles of expert's independence, objectiveness, comprehensiveness and completeness of investigations using modern achievements of science and technology.

4.2 Expert's independence. During expertise conduct, an expert community member appears independent; actually, an expert might not depend on the authority or person initiating expertise or other parties and persons interested in expertise results. An expert reports his opinion on the strength of analysis according to his special knowledge and awareness. Any persons found guilty of influencing an expert bear responsibility according to the legislature of the country.

4.3 Objectiveness, comprehensiveness and completeness of investigations. An expert performs investigations on a strictly scientific and practical basis, within his professional competence, in the objective, comprehensive and complete form.

Expert's report must proceed from propositions which admit validity/authenticity verification based on generally accepted scientific and practical data.

5. Rights and duties of expert community members.

5.1 An expert community member has the right for:

- the equal level of accessing all discussed materials with other members;
- the free expression of his civil and professional positions and expert assessments on all discussed subjects;
- the objective consideration of his viewpoint and expert assessments equally with the viewpoints and expert assessments of other members;
- remuneration of his work (under an employment agreement);
- the objective consideration of his expert activity during preparation of scientific reports (for researchers and tutors);
- personal data protection.

5.2 An expert community member has the following duties:

- regular participation (at least, once a month) in the activity of an expert community by registration on its official website, analysis, discussion and expert assessment of all materials published on the website;
- participation in electronic brainstorming, networked strategic conversations according to regulations (by agreement);
- commentation and preparation of answers on official questions;
- well-timed examination of e-mail messages received by an expert community on a specific subject domain within his certified competence;
- adherence to the ethical norms of discussions, in accordance with Professional Ethics Code of an expert, which implies:

  - correct consideration of discussion issues (instead of discussants and other issues beyond the scope of discussion);
  - adherence to weighted and responsible judgments and assessments, avoidance of far too emotional (or even pejorative and insulting) expressions, abusive language, etc.

5.3 Expert community members demonstrating neglectful attitude towards their duties receive an official warning. Subsequent breach of duties may lead to membership cancelation.

6. Motivation and remuneration of expert community members.

6.1 The following factors stimulate the activity of an expert:

- participation in managerial decision-making;
- personal reputation growth;
- intellectual property protection;
- expert activity assessment via participation in rating assignment;
- acquisition of exclusive analytical information;
- consideration of expertise as scientific research activity;
- usage of analytical and computational services;

- accreditation, certification, decorations;
- remunerations (on a contractual basis).

6.2 Generally, the assessments and decisions of an expert community on discussed documents have the form of recommendations. Exceptions are contracts for detailed expertise of certain issues, when experts and expert organizations bear a well-defined responsibility for possible losses.

7. Informational communication.

7.1 The official website (portal) of an expert community serves for studying documents submitted for expertise, organizing discussions between expert community members, publishing supplementary materials and comments on discussed issues.

7.2 The entire content of the portal is available for full members of an expert community only. Any user with Internet access may study selected materials.

7.3 Only full members of an expert community have access to publish their opinions and materials on the portal and to participate in voting.

7.4 Expert procedures, discussions on each document submitted for expertise are implemented either on dedicated forums of the portal, or according to established regulations involving moderators, in particular,

- experts' commenting of queries;
- assessment of events by semantic scales;
- electronic brainstorming;
- networked strategic conversation or discussion.

In the case of the first and second procedures, experts receive questionnaires with estimation scales. Empty forms of questionnaires are sent to e-mails specified by an expert during his registration on the website; the same rule applies to filled forms. Otherwise, they are eliminated from consideration. The third and fourth procedures correspond to moderation of experts' communication according to certain regulations.

7.5 Summing-up the discussion of a subject can employ "voting" on the website. For this, a moderator defines and announces voting rules and a corresponding period of time.
During voting on an issue, each member of an expert community disposes of one vote only. A moderator can introduce correction factors for assessments of different experts based on their ratings and competencies. Second votes of experts on a same issue are ignored.

7.6 Contracts for expert services are concluded in the case of detailed expertise of issues, documents and settlement of problem situations.

7.7 Final reports of expertise are published on the portal of an expert community for full members exclusively.

8. Activity organization.

8.1 For activity organization, it is possible to elect a representative body of a networked expert community by simple majority of votes (e.g., an organization committee, council of expertise, etc.).

The representative body with the status of a legal entity makes decisions on creating or choosing another legal entity to defend the interests of the community including conclusion of expertise contracts.

8.2 The representative body of an expert community consists of two primary departments. The first department focuses on the subject-oriented and methodological activity. The members of this department can be called subject supervisors. They have strong competence in problem solving, formulate tasks and specify the initial query to experts, as well as design methods, develop theory and improve technologies of expert procedures. The second department fulfils organizational functions, the members of this department can be called moderators. They perform administration and control, contact expert communities and professional communities, as well as organize concrete expert procedures. This department works directly with experts, i.e., sends questions to them, controls answers and responses and feedback of experts.

8.3 Regular methodological conversations, seminars and conferences are conducted among the members of the representative body (at least, once in 6 months).

8.4 General communication between an expert community and a client during expertise and expert activity support can be provided by a dedicated operator.

8.5 The operator may perform:

- design, hosting and support of the official website of an Expert community;
- registration of expert community members and organization of their work on the official website;
- organization of moderation services for expert procedures;
- following up the execution of duties by expert community members;
- identification of most active and professional experts, conclusion of agreements with them and organization of in-depth expertise;
- computer simulation of problem situations;
- rating assignment for experts and expert organizations;
- financing of works including contracting organizations and experts;
- preparation of accounts.

8.6 Being the moderator, the operator may fulfill the following duties:

- to conduct ("administer") expertise;
- to send queries to experts;
- to perform modeling of problem situations;
- to inform an expert about queries sent to him;

- to give necessary explanations to an expert;
- to control acquisition of experts' answers;
- to define the completion of different stages in expert procedures;
- to guarantee information security.

Being the moderator, the operator may enjoy the following rights:

- to prepare proposals in draft projects and decisions of the client;
- to prepare proposals on the development and financial provision of expertise support tools;
- to add new experts in mailing lists of queries;
- to determine the degree of urgency of expert procedures;
- to resend queries to other moderators for expertise on another subject;
- to suggest candidates for inclusion in the register of experts to the client.

A.2 Typical Regulations of Networked Expert Community Formation Within the Structure/for the Demands of Federal Authorities

1. Networked expert communities within the structure/for the demands of federal authorities (hereinafter referred to as expert communities) can be formed by appropriate legal and normative acts on the proposal of official heads of departments in a federal authority taking into account written proposals by the managers of corporations and organizations, representatives of civil society institutions, or separate citizens.

2. An expert community is formed within 6 months since the date of consent by the head of a federal authority. This means elaboration and activation of the following elements:

- an accreditation procedure for networked experts, organizations and communities;
- a portal for implementation of networked expert procedures, maintenance of the registers of experts, expert organizations and communities;
- regulations and techniques for implementation of networked expert procedures and maintenance of registers;
- a Professional Ethics Code for networked experts;
- an analytical treatment technology for expert information, an expert support technology for decision-making processes and, if necessary, other normative documents and techniques.

3. The Statute of an expert community interacting with a federal authority and its staff is established by a legal act of this authority.

The Statute of an expert community defines its goals, tasks, competence, organizational and financial provision, as well as the regulations of its activity, expertise conduct, formation, development and public participation in its work during expertise.

4. Expert community members fulfill their duties on a voluntary basis or, under in-depth examination of issues, according to contracts.

5. A federal authority initiating expert community formation implements the organizational and technical provision of the expert community according to an established legal procedure including products' supply, performance of work and rendering of services for state demands.

A.3 Assessment Criteria for Expert Analysis Organisations (Think Tanks)[1]

1. Direct relationship between Think Tank's efforts in a particular area, and positive change in that area, i.e., the amount of goods and services available to citizens, the state of physical and mental health, the quality of environment, the extent of political rights, access to institutions.

2. Growing up the quality level of services: improvement of people's satisfaction with Think Tank's services and permanent costs reduction of the services.

3. Extent to which respected publishers agree to publish Think Tank's works, especially on a nonprofit basis.

4. Ability to retain elite scholars and analysts.

5. Access to elites in the area of policymaking, media, and academia.

6. Academic reputation (formal accreditation; citations; publication in major academic books, journals, conferences, etc.).

7. Media reputation (the number of media appearances, interviews, and citations).

8. Reputation with policymakers (name recognition with particular issues, the number of briefings and official appointments, policy briefs, legislative testimony delivered).

9. The level of Think Tank's financial resources (endowment, membership fees, annual donations, government and private contracts, earned income).

10. The ability of a Think Tank to meet the demands of those that fund it or to meet the goals of its respective grant-making institution.

11. The overall output of a Think Tank (policy proposals, publications, interviews, conferences, staff nominated to official posts).

12. The number of recommendations adopted by policymakers, staff serving advisory roles to policymakers, awards given to scholars.

13. The usefulness of Think Tank's information in advocacy work, preparing legislation or testimony, preparing academic papers or presentations, conducting research, or teaching.

---

[1] This section employs criteria suggested by The University of Pennsylvania (USA) for rating assignment in think tanks (2009).

14. Think Tank's ability to produce new knowledge or alternative ideas on policy.
15. Ability to bridge the knowledge gap between government officials, business representatives and citizens.
16. Capacity for innovation and creativity.
17. Think Tank's ability to participate in social, professional, expert networks.
18. Quality assessment of expert materials by clients and users.

A.4 Assessment Procedure for Expert Ratings

In the first place, the quality of networked expertise depends on the competence of experts, their ability to appraise a current situation, to identify important factors and their correlation. Expert rating systems serve for assessing the activity of experts.

By a generally accepted assumption, the highest priority belongs to the opinions of specialists in a dedicated subject domain. At the same time, the staff of an expert group may include specialists in other subject domains. Anyway, it is necessary to define the relative weights of experts in assessing different groups of indicators.

To construct an expert rating system, one should acquire information on expert activity and perform automatic rating assignment. This requires acquiring data on the following groups of indicators (criteria):

- *individual characteristics of experts*, which reflect their motivation, wishes, interests, and needs;
- the *activity parameters of experts*, which reflect their current expert activity including collective activity;
- the *competence assessment parameters of experts* (professionalism).

Table A.1 combines several assessment criteria and indicators of expert activity, indicating of the feasibility of their automatic or expert evaluation. This table reproduces the absolute case with sufficient rating assignment resources. In practice, one can simplify the list of indicators.

To obtain the integrated rating of an expert using the criteria and indicators from the table, it is possible to apply the method of hierarchy analysis and paired comparison. The hierarchy of expert's characteristics has four levels: (1) "Expert's quality" integrated rating, (2) the three criteria from Table A.1, (3) indicators corresponding to each criterion above (see Table A.1), and (4) experts proper.

The procedure of paired comparison is executed by experts. For each pair of components on a same level of the hierarchy, they define how stronger is the influence of the first component on a higher component than that of the second one (e.g., in some estimation scale).

To reduce the intensiveness of integrated rating evaluation, one can eliminate level 3 from the hierarchy. In this case, most important criteria are moved to level 2. For instance, level 2 can be supplemented by the following criteria: "Creativity," "Independence," "Activity," "Experience," and "Professionalism."

Table A.1 Assessment criteria and indicators, the feasibility of their automatic computation or expert evaluation

Criteria and Indicators	Evalution
1. *Individual characteristics*	By expertise
1.1. Responsibility for expert activity	By expertise
1.2. Creativity (creative approach)	By expertise
1.3. The level of satisfaction from active expert actions and problem solving	By expertise
1.4. Pleasure from collective expert activity, positivity of expert position	By expertise
1.5. The level of (administrative and economic) independence	By expertise
1.6. The level of self-sufficiency, preference to in-depth and sympathetic understanding of colleagues	By expertise
1.7. Susceptibility and adaptation to an external environment (market dynamics)	By expertise
1.8. The characteristics of expert's emotional potential evaluated by the EQ technique (see [29])	By expertise
1.9. The characteristics of expert's intelligence potential evaluated by the IQ technique	By expertise
2. *The activity parameters of experts*	By expertise
2.1. The number of subject domains an expert specializes on (expert's mental outlook)	Automatic
2.2. The frequencies of queries to an expert and characteristics of his answers	Automatic
2.3. The correspondence between the materials queried from an expert and provided by him	Automatic
2.4. The period of expert's answers	Automatic
2.5. The fullness of expert's profile and individual characteristics of experts	Automatic
2.6. The dynamics of expert's communication with colleagues during expertise	Automatic
2.7. The completeness and accuracy of answers to queries	By expertise
2.8. The efficiency characteristics of expert activity evaluated by the number of expert's texts used in official normative documents, conceptual models and decisions made	Automated
2.9. The assessment of expert's activity by an ultimate user, subject supervisor, moderator	By expertise
2.10. The weighted assessment of expert's activity by other experts	By expertise
3. *The competence assessment parameters of experts*	By expertise
3.1. The number of scientific publications on a subject	Automatic
3.2. Practical experience in a subject domain	Automatic
3.3. Education, academic degree, academic rank	Automatic
3.4. Experience in expertise conduct including scientific and technical expertise (in the last 5 years)	Automated
3.5. Participation in Academic Councils, Scientific and Technical Councils, Thesis Boards	Automatic
3.6. Additional information	By expertise
3.7. Recommended experts	Automatic
3.8. The characteristics of intellectual property created and registered during expert activity	Automated

A.5 Professional Ethics Code for Networked Expert Communities

Professional Ethics Code for networked expert communities bases on international norms and national legal acts, on universal moral standards of behavior by state officials, representatives of business companies and society members, as well as on the Codes of conduct of expert communities in different areas of activity.

Article 1. Scope of the Code. The Code represents a set of general ethical principles and primary rules of behavior of expert community members (hereinafter referred to as experts): expert organizations, expert communities and separate experts. Prior to accreditation as an expert, a candidate gets acquainted with regulations of the Code and engages for following them during his expert activity.

Article 2. Purpose of the Code. The Code establishes the ethical norms and rules for the respectable behavior of experts during their activity, strengthens the reputation of experts, creates the trust space among state officials, representatives of business companies and society, as well as to provide a uniform moral and ethical foundation of expert activity.

The Code serves for outlining moral norms in expert activity and makes a public control tool for the ethical aspects of expert activity.

The knowledge and fulfillment of the Code by an expert is a key assessment criterion of his activity.

Article 3. Mission of experts. The Code is intended to improve the efficiency of expert activity in different subject domains. Fitting in the field of expert activity and recognizing the responsibility to the country, society and citizens, experts have to:

- perform their activity on a high professional level, be independent from citizens, professional or social groups, companies and organizations;
- eliminate actions preventing from conscientious expert activity, to remain neutral, to follow the norms of professional ethics and rules of business communication;
- show tolerance and respect to the local customs and traditions, consider the cultural and other features of different ethnic or social groups, confessions, promote cross-national and interconfessional consent;
- refrain from any behavior causing the slightest doubt in objective expertise, take measures to avoid conflicts of interests;
- refrain from using his reputation and status of an accredited expert to influence state authorities, organizations, officials for personal goals;
- keep immaculate appearance to promote the respect of citizens, representatives of business companies, local and federal authorities and to agree with business style;
- make public announcements, judgments and assessments within the limits of professional ethics.

Article 4. Principles of professional activity. An expert considers his membership in an expert community as a certain way of achieving many goals unattainable for him in the case of independent activity. Establishment of ethical norms (principles) becomes an inherent attribute of expert activity improvement. During organization of their activity, experts are guided by the following principles:

- *Honesty.* Everybody expects honest, holistic and fair appraisals from experts;
- *Responsibility.* Experts recognize their responsibility to society, have positive intentions and perform correct actions satisfying all interested sides and citizens affected by expertise;
- *Competence.* An expert has to keep improving his level of competence. This process should be accompanied by the development of expert's motivation, ideas and intentions;
- *Avoidance of harm.* Correct intentions and actions of an expert, at least, do no harm to people. Experts engage for applying decision-making methods which cause no damage to the country and society;
- *Triunity.* When solving a specific problem, an expert necessarily takes into account, at least, three major aspects of reality: the completeness of consideration and connection of the problem and its environment (*integrity, holistic*), the permanent and chaotic changes in a current situation (*variability*), and logical conditionality of some component of the problem (*order*);
- *Analyticity.* An expert has to analyze the set of interconnected issues such as improvement of the quality and standards of living, the growth of people's satisfaction with services, reduction of production costs, war against poverty, the stable development of the national economy, the cultural diversity of society, life safety, and ecological responsibility. An expert has to differentiate between verified facts and doubtful assumptions, to characterize a problem situation in the descriptive form;
- *Leadership.* Expert decisions aim at creating the atmosphere of leadership in organizations and an expert community;
- *Impartiality.* While making decisions, an expert is the state of composure and concentration. The feeling of anger, envy and fear are alien to an expert. While making decisions, he is attentive and controls the impact of negative emotions;
- *Adequacy.* An expert observes current events in their true colors; he is able to analyze possible consequences according to the viewpoints of other people. After decision-making, an expert asks himself the following question. Are the consequences of my decision beneficial to the interested organization and other people it affects? Answering this question, he imagines the interests of other people;
- *Process Approach.* While identifying the causes and effects of certain events, an expert gets focused on the corresponding process rather than these causes and effects. The process can be multiple-factor, illogical, nonlinear, and intermittent or have implicit evidence.

An expert disposes of:

- the self-control system for negative emotions (including an "early prediction subsystem" to diagnose consciousness recovery at early stages), e.g., invalid critics; inadmissibility of manipulation, strategy-proof expertise procedures;
- the facilities of informational society including network technologies of trust space formation, strategic consent achievement on the goals and ways of actions, collective expert procedures for decision-making.

Article 5. Business communication. During communication, an expert follows constitutional guarantees that a citizen, his rights and liberties are the supreme value and each citizen has the right for privacy, personal and family secret, protection of honor, dignity and good reputation.

During communication with employees, officials, citizens and colleagues, it is totally inadmissible to express:

- any form of discrimination (sex, age, race, nationality, language, citizenship, social/property/marital status, political or religious convictions);
- arrogance, rude manner, self-conceit, incorrectness of remarks, wrongful or undeserved incriminations;
- threats, outrageous statements or replicas.

Experts have to facilitate constructive collaboration in an expert community.

Article 6. Expert's responsibility for violation of the Code. An expert bears moral responsibility for violation of the Code and can be excluded from an expert community.

A.6 Security Problems

Technological expansion has a gradually increasing pace, and sometimes society becomes unable to comprehend new capabilities and threats generated by a certain technology. For instance, the discovery of atomic energy was followed by a clear understanding of scientists about possible risks of military application (recall the well-known *Letter from A. Einstein to USA President F.D. Roosevelt about the possible construction of nuclear bombs written in* 1939). However, today even specialists in information and communication technologies do not completely comprehend the social impact of the latter. No doubt, information and communication technologies provide numerous capabilities for decision-making, particularly, for expert activity. On the other hand, new problems arise.

The results of networked expertise including the ones obtained via modern information and communication technologies are adopted in decision-making. This sharpens the issues of *security*, i.e., protection of decisions against the negative impact of different participants of expert activity. Certain aspects of security problems (*viz.*, expertise manipulation effects) have been studied in Chap. 5 of the book.

In addition, society and state structures get interested in online networks as sources of specific information to reveal nascent tendencies that can be influenced. Such networks also represent a powerful means of information propagation, leading to the problem of data flows control.

In other words, there arise different problems of information security, both at the levels of individuals and society or state. Social and expert networks are actually potential scene of informational contagion, where control subjects struggle for the minds of other network participants.

Nowadays, it is possible to outline the following triad of situation development levels in a network: (1) informational interaction of participants, (2) informational impacts as a method of networked control, and (3) informational contagion. This forms a new challenge for control theory. The most important practical result of such research lies in recognizing that a social or expert network can be a tool and object of informational control and informational contagion. Presently, a few experts or participants of social networks think about the fact that they can be an object of informational control. A few subjects responsible for information security think about the fact that a social network can be an object or tool of informational impact.

The issues of information safety gain particular importance in the context of networked expertise development. Generally, experts implement representation mechanisms in decision-making within state authorities and business companies. They represent the interests of certain layers of state authorities, business communities and, of course, their own interests. Corporations and state authorities have to consider the opinion of experts. This follows from business needs and the existing normative basis of decision-making.

References

1. Aizerman, M., Aleskerov, F.: Theory of Choice. Amsterdam, Elsevier (1995)
2. Abramova, N., Avdeeva, Z., Kovriga, S., Makarenko, D.: Subject-formal methods based on cognitive maps and the problem of risk due to the human factor/cognitive maps. In: Perusich, K. (ed.) Vienna: In Tech, pp. 35–63 (2009)
3. Ackerman, M., McDonald, D., Lutters, W., Muramatsu, J.: Recommenders for expertise management. In: Proceedings of the ACM SIGIR workshop on recommender systems: algorithms and evaluation (SIGIR'99). University of California, Berkeley (1999)
4. Aerts, D., Czachor, M.: Quantum aspects of semantic analysis and symbolic artificial intelligence. J. Phys. A: Math. Gen. **37**, L123–L132 (2004)
5. Aldrich, J., Alt, J.: Positive changes in political science: the legacy of Richard D. McKelvey's most influential writings. Michigan, University of Michigan Press (2007)
6. Aleskerov, F.: Power indices taking into account agents' preferences/mathematics and democracy. In: Simeone, B., Pukelsheim, F. (eds.) Mathematics and Democracy, pp. 1–18. Berlin, Springer (2006)
7. Ali, P., Gregoriou, G.: International corporate governance after Sarbanes-Oxley. London, John Wiley and Sons (2006)
8. http://answers.google.com/answers
9. Arrow, K., Sen, A., Suzumura, K.: Handbook of social choice and welfare. Gulf Professional Publishing, New York (2002)
10. Avdeeva, Z., Kovriga, S.: Cognitive approach in simulation and control. In: Proceedings of the 17th IFAC, pp. 1613–1620. World Congress, Seoul, South Korea (2008)
11. Aumann, R., Myerson, R.: Endogenous formation of links between players and coalitions: an application of the shapley value/the shapley value, In: Roth, A. (ed.) The Shapley Value, pp. 175–191. Cambridge University Press, Cambridge (1988)
12. Axelrod, R., Conflict of interest: a theory of divergent goals with applications to politics. Chicago, Markham Publ. Comp. (1970)
13. Balog, K., Azzopardi, L., De Rijke, M.: Formal models for expert finding in enterprise corpora. In: Proceedings of the 29th Annual International Acm Sigir Conference on Research and Development in Information Retrieval (SIGIR'06), pp. 43–50. ACM, New York (2006)
14. Balog, K., de Rijke, M.: Determining expert profiles (with an application to expert finding). In: Proceedings of the 20th International Joint Conference on Artificial Intelligence, pp. 2657–2662. Morgan Kaufmann, San Francisco (2007)
15. Balog, K., de Rijke, M.: Non-local evidence for expert finding. In: Proceedings of the 17th ACM Conference on Information and Knowledge Management, pp. 489–498. ACM, New York (2008)
16. Balog, K., Weerkamp, W., de Rijke, M.: A few examples go a long way: constructing query models from elaborate query formulations. In: Proceedings of the 31th Annual International ACM SIGIR Conference on Research and Development in Information Retrieval, pp. 371–378 (2008)

17. Banks, R.: Sophisticated voting outcomes and agenda control. Soc. Choice Welfare **1**(4), 295–306 (1985)
18. Barberà, S., Berga, D., Moreno, B.: Individual versus group strategy-proofness: when do they coincide. J. Econ. Theory **145**(5), 1648–1674 (2010)
19. Barberà, S.: Strategy-proof social choice. In: Arrow, K., Sen, A., Suzumura, K. (eds.) Handbook of Social Choice and Welfare, vol. 2, pp. 242–275 (2006)
20. Bloch, F., Jackson, M.: Definitions of equilibrium in network formation games. Int. J. Game Theory **34**(3), 305–318 (2006)
21. Breslin, J., Bojars, U., Aleman-Meza, B., Boley, H., Mochol, M., Nixon, L., Polleres, A., Zhdanova, A.: Finding experts using internet-based discussions in online communities and associated social networks. First International ExpertFinder Workshop, pp. 38–47 (2007)
22. Bogomolnaina, A., Jackson, M.: The stability of hedonic coalition structures. Games Econ. Behav. **38**, 201–230 (2002)
23. Bossert, W., Weymark, J.: Social choice: recent developments/the new palgrave dictionary of economics, 2nd edn. Palgrave Macmillan, London (2006)
24. Burkov, V., Goubko, M., Korgin, N., Novikov, D.: Mechanisms of organizational behavior control: a survey. Adv. Syst. Sci. Appl. **13**(1), 1–13 (2013)
25. Burkov, V., Goubko, M., Korgin, N., Novikov, D.: Integrated mechanisms of organizational behavior control. Adv. Syst. Sci. Appl. **13**(2), 1–9 (2013)
26. Burkov, V., Goubko, M., Kondrat'ev, V., Korgin, N., Novikov, D.: Mechanism design and management: mathematical methods for smart organizations. In: Novikov, W. (ed.), p. 163. Nova Science Publishers, New York (2013)
27. Campbell, C., Maglio, P., Cozzi, A., Dom, B.: Expertise identification using E-mail communications. In: Proceedings of CIKM'03, pp. 528–531. ACM, New York (2003)
28. Craswell, N., Hawking, D., Vercoustre, A., Wilkins, P.: Panoptic expert: searching for experts not just for documents. In: Ausweb Poster Proceedings. Queensland, Australia (2001)
29. Cooper, R.K., Sawaf, A.: Executive EQ. In: Emotional Intelligence in Business, pp. 358. Texere, New York (2000)
30. Dan, A.: Predictably irrational: the hidden forces that shape our decisions. HarperCollins, New York (2008)
31. Garratt, B. (ed.): Developing Strategic Thought. A Collection of the Best Thinking on Business Strategy. Profile Books, Hong Kong (2003)
32. Diederich, J., Iofciu, T.: Finding communities of practice from user profiles based on folksonomies. In: Proceedings of the 1st International Workshop on Building Technology Enhanced Learning Solutions for Communities of Practice (TEL-CoPs'06). Crete (2006)
33. Druker, P.: Practice of Management. Perennial Library, New York (1986) (Electronic books)
34. Erev, I., Haruvy, E.: Learning and the economics of small decisions. In: Handbook of Experimental Economics Results (2013). http://www.utdallas.edu/~eeh017200/papers/LearningChapter.pdf
35. ExpertFinder. http://expertfinder.info/, http://wiki.foaf-project.org/w/. Accessed 15 Nov 2010
36. Firestone, J., McElroy, M.: Doing knowledge management. Learn. Organ. J. **12**(2), 189–212 (2005)
37. Friedman, A., Miles, S.: Stakeholders: Theory and Practice. Oxford University Press, Oxford (2006)
38. http://www.foaf-project.org
39. Gavrilova, M.: Computational intelligence: a geometry-based approach, In: Kacprzyk, J. (ed.) Springer Engineering book series Studies in Computational Intelligence. Springer-Verlag (2009)
40. Gavrilova, M., Monwar, M.: Multimodal Biometrics and Intelligent Image Processing for Security Systems, IGA book (2013)

41. Granovetter, M.: Threshold models of collective behavior. AJS **83**(6), 1420–1443 (1978)
42. Groves, T.: Efficient collective choice when compensation is possible. Rev. Econ. Stud. **46**(2), 227–241 (1979)
43. Gubanov, D., Korgin, N., Novikov, D.: Models of reputation dynamics in expertise by social networks. In: Proceedings of the UKACC International Conference on Control, pp. 203–210. Coventry University, Coventry (2010)
44. Gubanov, D., Korgin, N., Novikov, D.: Network expertise and dynamics of reputation. In: Proceedings of the 10th International Meeting of the Society for Social Choice and Welfare, p. 27. HSE, Moscow (2010)
45. Guriev, S.: Red tape and corruption. J. Dev. Econ. **73**(2), 489–504 (2004)
46. Intelligent Quantum & Soft Computing R&D Group.: http://www.qcoptimizer.com
47. Iofciu, T., Diederich, J., Dolog, P., Balke, W.: ExpertFOAF recommends experts. In: First International ExpertFinder Workshop. Berlin (2007)
48. Jackson, M.: A crash course in implementation theory. Soc. Choice Welfare **18**, 74–83 (2001)
49. Jackson, M.: Mechanism Theory. The Encyclopedia of Life Support Systems (2000)
50. Jackson, M.: Social and Economic Networks. Princeton Univ. Press, Princeton (2010)
51. Jackson, M.: The economics of social networks. In: Blundell, R., Newey, W., Persson, T. (eds.) Advances in Economics and Econometrics, Theory and Applications: The 9th World Congress of the Econometric Society, vol. 1, pp. 1–56. Cambridge University Press, Cambridge (2006)
52. Park, J.J., Leung, V.C.M., Shon, T., Wang, C.: Future Information Technology, Application, and Service: FutureTech. vol. 1, p. 823. Springer (2012)
53. Kaplan, R.S., Norton, D.P.: Strategy Maps: Converting Intangible Assets into Tangible Outcomes. Harvard Business School Publishing Corporation, Harvard (2005)
54. Korgin, N.: Use of intersection property for analysis of feasibility of multicriteria expertise results. Autom. Remote Control. **71**(6), 1169–1183 (2010)
55. Korgin, N., Burkov, V., Iskakov, M.: Application of generalized median voter schemes to designing strategy-proof mechanisms of multicriteria active expertise. Autom. Remote Control. **71**(8), 1681–1694 (2010)
56. Korgin, N., Burkov, V., and Iskakov, M.: On strategy-proof direct mechanism of active expertise over strictly convex compact set. Autom. Remote Control. **71**(10), 2168–2175 (2010)
57. Kleinberg, J.: Authoritative sources in a hyperlinked environment. In: Proceedings of the 9th ACM SIAM Symposium on Discrete Algorithms, pp. 604–632 (1998)
58. Klimenko, S., Raikov, A.: Electronic brainstorming. In: Proceedings of the International Scientific-Practical Conference Expert Community Organization in the Field of Education, Science and Technologies, pp. 181–185. Triest, Italy. Sept 26–27 2013
59. Lappas, T., Liu, K., Terzi, E.: Finding a team of experts in social networks. In: Proceedings of the 15th ACM SIGKDD International Conference on Knowledge Discovery and Data Mining, pp. 467–476. ACM, New York (2009)
60. Le Breton, M., Zaporozhets, V.: On the equivalence of coalitional and individual strategy-proofness properties. Soc. Choice Welfare **33**(2), 287–309 (2008)
61. Lee, S.F.: Building Balanced Scorecard with SWOT analysis and implementing sun tzu's the art of business management strategies on qfd methodology. Manag. Auditing J. **15**(1), 68–76 (2000)
62. Litvintseva, L., Ulyanov, S.: Intelligent control system/Quantum computing and self-organization algorithm. J. Comput. Syst. Sci. Intern. **48**(6), 946–984 (2009)
63. Li, J., Boley, H., Virendrakumar, C., Mei, J.: Expert finding for ecollaboration using foaf with ruleml rules. In: Proceedings of the Montreal Conference on eTechnologies (MCeTech) 2006
64. www.livejournal.ru

65. Lin, C., Yang, J., Cai, R., Wang, X., Wang, W., Zhang, L: Modeling semantics and structure of discussion threads. In: Proceedings of the 18th International Conference on World Wide Web (WWW'09). ACM Press, New York (2009)
66. Lindner, T., Nehring, K., Puppe, C.: Which voting rule is more manipulatable? results from simulation studies. In: Proceedings of the 10th International Meeting of the Society for Social Choice and Welfare, p. 14. HSE, Moscow (2010)
67. Litvak, B.G.: Expert technologies in management. Delo, Moscow (2004) (in Russian)
68. Liu, X., Croft, W.B., Koll, M.: Finding experts in community-based question-answering services. In: Proceedings of the CIKM ACM International Conference on Information and Knowledge Management (CIKM'05), pp. 315–316. ACM Press, New York (2005)
69. Lu, W., Robertson, S., Macfarlane, A., Zhao, H.: Window-based enterprise expert search. In: Proceedings of the 15th Text Retrieval Conference, 2006
70. Macdonald, C., White, R.: Usefulness of click-through data in expert search. In: Proceedings of SIGIR'09. Boston (2009)
71. McKelvey, R.: A theory of optimal agenda design. Manag. Sci. 27(3), 303–321 (1981)
72. Millen, R., Feinberg, J., Kerr, B.: Dogear: social bookmarking in the enterprise. In: Proceedings of the Conference on Human Factors in Computing Systems, pp. 172–179. Montreal (2006)
73. Mintsberg, H., Ahlstand, B., Lampel, J.: Strategic safary. A Guided Tour through the Wild of Strategic Management. Prentice Hall, New York (1998)
74. Moulin, H.: On Strategy-proofness and single-peakedness. Public Choice 35(4), 437–455 (1980)
75. Mueller, D.: Public Choice III. Cambridge University Press, Cambridge (2003)
76. Myerson, R.: Game Theory: Analysis of Conflict. Harvard University Press, Cambridge (1997)
77. Novikov, A., Novikov, D.: Research Methodology: From Philosophy of Science to Research Design, p. 130. CRC Press, Leiden (2013)
78. Novikov, D., Chkhartishvili, A.: Reflexion and Control: Mathematical Models, p. 298. CRC Press, Leiden (2014)
79. Novikov, D.: Control Methodology, p. 101. Nova Science Publishers, New York (2013)
80. Novikov, D.A.: Theory of Control in Organizations, p. 341. Nova Science Publishing, New York (2013)
81. Orlov, A.I.: Decision-Making Theory. Examen, Moscow (2005) (in Russian)
82. http://otvety.google.ru/otvety
83. Page, L., Brin, S., Motwani, R., Winograd, T.: The PageRank Citation Ranking: Bringing Order to the Web. Stanford Digital Libraries Working Paper (1998)
84. Paul, PP., Gavrilova, M., Klimenko, S.: Situation Awareness of Cancelable Biometric System, The Visual Computer, Springer. Accepted (2013)
85. Peleg, B.: An axiomatization of the core of cooperative games without side payments. J. Math. Econ. 14(2), 203–214 (1985)
86. Penrose, L.: On the Objective Study of Crowd Behavior. H. K. Lewis & Co., London (1952)
87. Peters, H.: Cooperative Games with Transferable Utility. In: Game Theory, vol. 1, pp. 121–131. Springer (2008)
88. Raikov, A.: Convergent cognitype for speeding-up the strategic conversation. In: Proceedings of the 17th IFAC World Congress, pp. 8103–8108. Seoul (2008)
89. Raikov, A. Convergent Control and Decisions Support. Ikar, Moscow (2009) (in Russian)
90. Raikov, A.N.: Holistic discourse in the network cognitive modeling. J. Math. Syst. Sci. 3(10), 519–530 (2013)
91. Raikov, A., Panfilov, S.: Convergent decision support system with genetic algorithms and cognitive simulation. In: Proceedings of the IFAC Conference on Manufacturing Modeling, Management and Control, (MIM'2013), pp. 1142–1147. St. Petersburg, Russia 19–21 June 2013
92. O'Reilly, T.: What is web 2.0. http://oreilly.com/web2/archive/what-is-web-20.html

93. http://profnet.prnewswire.com/PRNJ.aspx
94. Roberts, R.: Discrete Mathematical Models with Applications to Social, Biological and Environmental Problems. Prentice Hall, New Jersey (1976)
95. Rocchio, J.: Relevance Feedback in Information Retrieval. The SMART Retrieval System: Experiments in Automatic Document Processing. Prentice Hall, New York (1971)
96. Rose-Ackerman, S.: Corruption. In: Readings in Public Choice and Constitutional Political Economy, vol. V, pp. 551–566 (2008)
97. http://ruleml.org
98. http://www.sap.com/portugal/solutions/business-suite/erp/hcm/pdf/50053653.pdf
99. Saaty, T.L.: The Analytic Hierarchy Process. University of Pittsburg, Pittsburg (1988)
100. Schwartz, M., Wood, D.: Discovering shared interests among people using graph analysis of global electronic mail traffic. Commun. ACM **36**(8), 78–89 (1993)
101. Sidelnykov, Y.: System Analysis of Expert Forecasting. MAI, Moscow (2007). (in Russian)
102. Skopik, F., Truong, H., Dustdar, S., et. al.: Trust and Reputation Mining in Professional Electronic Communities, pp. 76–90. Springer, Heidelberg (2009)
103. Serdyukov, H., Rode, F., Hiemstra, D.: Modeling multi-step relevance propagation for expert finding. In: Proceedings of CIKM'08, pp. 1133–1142. ACM, New York (2008)
104. Seo, J., Croft, W.: Thread-based expert finding. In: Proceedings of the SIGIR SSM Workshop (SIGIR'09). SSM, Boston (2009)
105. Surowiecki, J.: The Wisdom of Crowds: Why the Many are Smarter than the Few and How Collective Wisdom Shapes Business, Economies, Societies and Nations. Doubleday, New York (2004)
106. Sihn, W., Heeren, F.: XPERTFINDER—Expert finding within specified subject areas through analysis of messages. In: Proceedings of the 1st International NAISO Congress on Autonomous Intelligent Systems. Geelong (2002)
107. http://sioc-project.org
108. Streeter, L., Lochbaum, K.: An expert/expert-locating system based on automatic representation of semantic structure. In: Proceedings of the 4th International Conference on Artificial Intelligence and Applications, pp. 345–350 (1988)
109. Takamiya, K.: Domains of social choice functions on which coalition strategy-proofness and maskin monotonicity are equivalent. Econ. Lett. **95**(3), 348–354 (2007)
110. Tirole, J.: A theory of collective reputations (with applications to the persistence of corruption and to firm quality). Rev. Econ. Stud. **63**(1), 1–22 (1996)
111. Ulyanov, S., Ghisi, F., Panfilov, S., Ulyanov, V., Kurawaki, I., Litvintseva, L.: Simulation of Quantum Algorithms on Classical Computers. Universita degli Studi di Milano, Polo Didattico e di Ricerca di Crema, Note del Polo, vol. 32, p. 96 (2000)
112. Ulyanov, S., Raikov, A.: Chaotic factor in intelligent information decision support systems. In: Aliev, R. et'al. The 3rd International Conference on Application of Fuzzy Systems and Soft Computing (ICAFS'98), pp. 240–245. Wiesbaden, Germany, 5–7 Oct 1998
113. Wasserman, S., Faust, K., Iacobucci, D.: Social Network Analysis: Theory and Methods. Cambridge University Press, Cambridge (1995)
114. http://www.w3.org/TR/skos-reference
115. Wong, P.C., Leung, L., Scott, M.J. et al.: Designing a collaborative visual analytics tool for social and technological change prediction. IEEE Comput. Graph. Appl. **29**(5), 58–68 (2009)
116. Yanushkevich, S., Gavrilova, M., Wang, P., Srihari S.: Image Pattern Recognition: Synthesis and Analysis in Biometrics, Book, Series in Machine Perception and Artificial Intelligence, vol. 67, p. 423. World Scientific Publishers (2007)
117. Yu, B., Singh, M.: Searching Social Networks. In: Proceedings of the AAMAS Workshop on Regulated Agent-based Social Systems: Theories and Applications (RASTA), 2002
118. Zhang, J., Ackerman, M., Adamic, L.: Expertise networks in online communities: structure and algorithms. In: Proceedings of the 16th International Conference on World Wide Web (WWW'07), pp. 221–230. ACM Press, New York (2007)

About the Authors

Dmitry Gubanov Cand. Sci. (Eng.), doctoral candidate at Trapeznikov Institute of Control Sciences of the Russian Academy of Sciences. Fields of research: informational control in social networks, knowledge representation, knowledge management.
E-mail: DimaGubanov@mail.ru.

Nikolai Korgin Cand. Sci. (Eng.), Assoc. Prof., leading researcher at Trapeznikov Institute of Control Sciences of the Russian Academy of Sciences, and associate professor at Moscow Institute of Physics and Technology. Fields of research: the theory of control in organizations, game theory, mechanism design.
E-mail: nkorgin@ipu.ru

Dmitry Novikov Dr. Sci. (Eng.), Prof., corresponding member of the Russian Academy of Sciences. Deputy Director of Trapeznikov Institute of Control Sciences of the Russian Academy of Sciences, and head of Control Sciences Department at Moscow Institute of Physics and Technology. Fields of research: the theory of control in interdisciplinary systems, system analysis, game theory, and decision-making. *E-mail*: novikov@ipu.ru.

Alexander Raikov Dr. Sci. (Eng.), Prof., full state councilor (class 3) of the Russian Federation, laureate of the Governmental Prize of the Russian Federation in the field of science and technology, director of analytical agency "New strategies," leading researcher at Trapeznikov Institute of Control Sciences of the Russian Academy of Sciences, and professor at the Russian Presidential Academy of National Economy and Public Administration, Moscow State Institute of Radio Engineering, Electronics and Automation, Higher School of Economics (National Research University). Fields of research: strategic management, artificial intelligence, expert analysis systems. *E-mail*: Alexander.N.Raikov@gmail.com.

Printed in the United States
By Bookmasters